Lyndon B. Johnson School of Public Affairs
Policy Research Project Report
Number 124

Colonia Housing and Infrastructure

Volume 2:
Water and Wastewater

A report by the
Colonia Housing and Infrastructure Policy Research Project
The University of Texas at Austin
1997

Library of Congress Card No.: 97-74582
ISBN: 0-89940-735-8
ISBN for three-volume set: 0-89940-737-4

Printed in the U.S.A.

Cover Design by Doug Marshall
LBJ School Publications Office

Policy Research Project Participants

Students

Laura Barrientos, B.A. (Psychology), The University of Texas at Austin

Gina M. Briley, B.A. (Mathematics/Political Science), Duke University

LaTanya Flix, B.A. (Sociology), Trinity University

Antonnette Gibbs, B.S. (Economics/Mathematics), University of Missouri-Columbia

Miguel A. Guajardo, B.S. (Education), The University of Texas at Austin

Alicia Hernandez Sanchez, B.S. (Civil Engineering), University of California at Berkeley

José Hernández, B.B.A., M.B.A. (Business Administration), Corpus Christi State University

Pamela Hormuth, B.A. (Psychology), The University of Texas at Austin

David Miller, B.S. (Education), Baylor University

Susan P. Salomone, B.S. (Foreign Service), Georgetown University

Michelle Salazar, B.A. (Plan II), The University of Texas at Austin

Alisha Sanders, B.A. (Sociology), Rice University

Barbara Simpson, B.S.W. (Social Work), East Texas State University

Christina Todd, B.A. (Political Science), University of Denver

Marialaura G. Valencia, B.A. (History), Columbia University

Project Directors

Jorge Chapa, M.A. (Demography), Ph.D. (Sociology), Stephen H. Spurr Centennial Fellow, Associate Professor of Public Affairs, Lyndon B. Johnson School of Public Affairs, The University of Texas at Austin

David Eaton, M.Sc. in Hygiene (Public Health), Ph.D. (Environmental Engineering), Bess Harris Jones Centennial Professor in Natural Resource Policy Studies, Lyndon B. Johnson School of Public Affairs, The University of Texas at Austin

Table of Contents

List of Tables

List of Figures

List of Acronyms

CDBG	Community Development Block Grant
BECC	Border Environment Cooperation Commission
CPLP	Colonia Plumbing Loan Program
CWTAP	Colonia Wastewater Treatment Assistance Program
COPC	Community Outreach Partnership Center
EDA	Economic Development Administration
EDAP	Economically Distressed Areas Programs
EPA	U.S. Environmental Protection Agency
FmHA	Farmers Home Administration (now RECDA)
HUD	U.S. Department of Housing and Urban Development
MSR	Model Subdivision Rules
NADBank	North American Development Bank
NAFTA	North American Free Trade Agreement
RECDA	Rural Economic and Community Development Agency (formerly FmHA)
STEP	Small Towns Environmental Program
TDHCA	Texas Department of Housing and Community Affairs
TWDB	Texas Water Development Board

Foreword

The Lyndon B. Johnson School of Public Affairs has established interdisciplinary research on policy problems as the core of its educational program. A major part of this program is the nine-month policy research project, in the course of which two or more faculty members from different disciplines direct the research of ten to thirty graduate students of diverse backgrounds on a policy issue of concern to a government or nonprofit agency. This "client orientation" brings the students face to face with administrators, legislators, and other officials active in the policy process and demonstrates that research in a policy environment demands special talents. It also illuminates the occasional difficulties of relating research findings to the world of political realities.

Colonias have been a key unresolved urban water resources problem in Texas for many years. The detailed understanding of the relationship between population and water infrastructure in colonias is of vital interest to Texas planners and policy makers. Two pressing colonia needs are (a) gaining access to water and wastewater infrastructure programs and (b) addressing the massively substandard housing conditions on the U.S. side of the Texas-Mexico border. Research for this report was made possible by a grant from the U.S. Department of Housing and Urban Development Community Outreach Partnership Center at The University of Texas at Austin (COPC-TX-95-0018). Financial support was also provided by the Stephen H. Spurr Fellowship.

The curriculum of the LBJ School is intended not only to develop effective public servants but also to produce research that will enlighten and inform those already engaged in the policy process. The project that resulted in this report has helped to accomplish the first task; it is our hope that the report itself will contribute to the second.

Finally, it should be noted that neither the LBJ School nor The University of Texas at Austin necessarily endorses the views or findings of this report.

Edwin Dorn
Dean

Acknowledgments and Disclaimer

The U.S. Department of Housing and Urban Development funded the research which forms the underlying basis for this report. The authors are solely responsible for the accuracy of the statements and interpretations contained in this publication. Such interpretations do not necessarily reflect the views of the government of the United States, the Lyndon B. Johnson School of Public Affairs, The Urban Issues Program, or The University of Texas at Austin.

We would like to thank Craig Pedersen, Executive Director of the Texas Water Development Board; Larry Paul Manley, Executive Director of the Texas Department of Housing and Community Affairs; and their staffs for their assistance in helping us obtain the data for this project. Thanks are also extended to Dan Torres, Special Assistant to the Texas Attorney General; Homer Cabello, Director of the Office of Colonia Initiatives at the Texas Department of Housing and Community Affairs; Professor Robert Wilson, LBJ School of Public Affairs; Amy Johnson; Vick Hines; John Hennenberger; Andy Homer; and Karen Paup. Much appreciation goes to Andy Alarcon, Lucy Neighbors, Martha Harrison, Brian Robinson, Martin Schulz, and Anadelia Romo, all of whom offered invaluable assistance with this project.

Executive Summary

This volume of the *Colonia Housing and Infrastructure* Policy Research Project Report examines state, federal, and international programs which provide financing for new or improved water distribution and wastewater collection and treatment infrastructure for the colonias in the Texas-Mexico border region. Much of the report focuses on the state programs, especially the Texas Water Development Board's (TWDB) Economically Distressed Areas Program (EDAP) and the Texas Department of Housing and Community Affairs' (TDHCA) Colonia Fund. The report identifies colonias which do not or will not have access to potable water and wastewater collection infrastructure even after the completion of the many projects currently underway. The report examines the cost estimate for providing additional infrastructure for these colonias and makes recommendations for the legislature to consider when making appropriations decisions.

The TWDB's EDAP program provides funding for both water and wastewater infrastructure, but does not allocate funds for individual homeowners' connections to the main lines. EDAP is funded by $250 million in general obligation bonds approved by Texas voters, as well as other state and federal sources. EDAP is the largest funding source, but can be difficult to access. Colonias must be represented by a political subdivision and therefore have little control over the decision to apply for funding, the design of the project, the area the project will serve, and the type of service to be implemented.

TDHCA's Colonia Fund programs provides funding for new or improved water and wastewater infrastructure, subdivision platting, street paving, street drainage, housing rehabilitation (including indoor plumbing and connection to main infrastructure), community centers, and neighborhood development and recreation. The Colonia Fund is financed primarily through a mandatory set-aside from the Texas state community development block grant allocation, providing nearly $9 million per year. Additional infrastructure and community development funds are available from the Texas Natural Resource Conservation Commission, the U.S. Department of Agriculture's Rural Economic and Community Development Agency, the U.S. Department of Commerce's Economic Development Administration, the U.S. Environmental Protection Agency, the Border Environment Cooperation Commission, and the North American Development Bank.

Current funding and future resource allocation for colonia infrastructure is dependent on estimates of both need and cost for infrastructure in the colonias. The TWDB is the only public agency which has completed and published an aggregate survey of the colonias' infrastructure needs, *Water for Texas: Water and Wastewater Needs of Colonias in Texas*. This survey, first conducted in 1992 and updated in 1995, has been instrumental in articulating the initial needs, providing the basis for justifying the investment in infrastructure, and assessing remaining needs. The survey is not perfect; the TWDB cautions that it is preliminary in scope. The methodology of the survey is generally sound, and the 1995 update of needs and cost estimates benefits tremendously from individual applications for infrastructure funding. This report concludes that the survey is primarily limited by the availability of personnel and time to accurately assess each of the 1,436 colonias identified.

This report examines in detail the extent to which current EDAP projects will meet the infrastructure needs of colonia residents. As of February 1996, EDAP had committed $384.5 million for 62 infrastructure projects in 831 colonias, affecting 209,716 residents. The figures computed in this report indicate that EDAP projects will serve less than the TWDB published figures; the report has calculated that nearly 59 percent of colonia residents and just over 56 percent of colonias will benefit from EDAP projects. The TWDB, on the other hand, stated that current projects would serve 75 percent of colonia residents and 65 percent of colonias. The difference reflects the fact that when calculating percentages the TWDB's accounting program failed to account for colonias which will be served by more than one EDAP project.

The TWDB has estimated that an additional $424.6 million is needed to provide both water infrastructure for 30,860 colonia residents and wastewater infrastructure for 139,626 colonia residents. Although there are funding sources other than the TWDB and EDAP, these sources combined are not expected to satisfy the deficit between available financing and needed infrastructure funding. Thus, it remains the responsibility of the State of Texas to determine the extent to which colonia infrastructure funding will be available. In making this determination, this report recommends the Texas Legislature consider:

- establishing statewide regulations that mandate utility provision of infrastructure prior to the subdivision, platting and sale of individual home lots;

- amending the EDAP legislation to allow individual colonias to make independent applications for infrastructure funding; and

- appropriating additional EDAP funding for colonias with unmet needs.

This report urges Texas to act quickly. Colonia populations are growing rapidly and a lack of potable water and wastewater removal infrastructure could become a barrier to the economic vitality of the entire border area.

Introduction

In 1992, the Texas Water Development Board (TWDB) completed the first aggregate survey of investment in water and wastewater infrastructure of Texas colonias, *Water and Wastewater Needs of Colonias in Texas*. The TWDB revised this infrastructure survey in 1995, recalculating colonia infrastructure requirements in 23 counties in the Texas-Mexico border region. In 1995, the TWDB counted 1,436 colonias which were home to an estimated total of 339,041 residents.[1] The update did not include colonias in all Texas counties, and the TWDB estimates that an additional 33 colonias exist in other counties in the border area.

The 1990s is the first decade in which the State of Texas has enacted and enforced laws aimed at halting the establishment and expansion of colonias. Thus far, Texas' approach to improving water and wastewater services to these substandard border communities has been to allow various agencies to administer diverse federal and state programs. Each program has as its goals the extension of service to colonias currently unconnected to water and wastewater systems, and the improvement of existing services to meet state and federal standards. This piecemeal approach has been difficult to coordinate. It has responded to many colonia needs, but also has left many unaddressed. The provision of infrastructure has proceeded at a pace that is slower than ideal, considering the health and environmental consequences of continued substandard service.

This report examines the current state of funding for colonia environmental infrastructure in Texas and the results of that investment. It first describes existing federal and state programs, as well as the potential for further assistance from new, NAFTA-related international sources. The report then analyzes the potential needs of colonias that may go unmet, in spite of the efforts of existing programs. The report also analyzes the methodology currently used by Texas to estimate colonia infrastructure needs, describing its strengths and shortcomings.

This report's recommendations amount to a set of value decisions that the State of Texas will have to make in order to address colonia infrastructure needs in a concerted and effective manner. The issue of colonia infrastructure is a microcosm of intergovernmental relations of many federal, state, and local institutions. At the end of a century in which Texas has seen phenomenal growth and economic development, many of its border residents (in many cases also its poorest residents) have yet to see running water in their homes. Few argue that colonia residents should continue to live with inadequate water and wastewater service, but the question of responsibility is less clear. Should counties be required to serve colonias? Should the state? Who bears the financial responsibility for cleaning up the results of unplanned and inadequately supported growth? Who bears the legal responsibility to ensure that existing colonias receive service and that new colonias cease to develop? If the answer to any of these questions is the State of Texas, border colonia policy will require increasing attention and resources in the next decade.

Appendix A contains tables for each EDAP-eligible county that list individual colonias and describe the scope and impact of each EDAP project. It is difficult to summarize succinctly the information contained in Appendix A. The tables were generated from the TWDB needs database in conjunction with the summary of EDAP projects as of February 15, 1996, paired with unpublished data provided by TWDB staff. Appendix A contains information on six additional counties not included in the 1995 needs survey update (Bee, Edwards, Jim Hogg, La Salle, Pecos, and San Patricio counties) and does not include Brewster County. Brewster County was not in the TWDB needs database, and the 1995 update located no colonias in Brewster County. Each table in Appendix A shows EDAP projects, by stage, with the colonias which will benefit. For projects affecting colonias served by previous EDAP projects, the colonia name is listed. To avoid double-counting, other data are not listed again, such as population, population not served by central water, population served and not served by central wastewater, number of dwellings, occupied lots, and total lots. Table 2 and Table 3 summarize some of the information contained in Appendix A.

Texas Colonia Development

Colonias are private property located outside city limits, some in nearby suburbs and others extremely remote. As unincorporated areas, colonias do not have local city governments. The Texas Water Development Board (TWDB) defines a colonia as a "primarily residential subdivision, with five or more housing units, characterized by inadequate water supply or wastewater removal infrastructures and lacking the financial resources needed to satisfy the need for water and/or wastewater infrastructures."[2]

The reasons for development of colonias along the border are twofold. First, many residents of border counties cannot afford even low-income housing in border cities. This is compounded by long waiting lists for subsidized housing in border cities, increasing demand for colonia lots even among families who can afford housing in incorporated areas. Second, developers have historically been allowed to sell land and structures without creating the necessary infrastructure to serve those who buy the land. Some also identify rapid population increases in Mexico and high immigration to the United States as reasons for colonia development.[3] But most colonia residents are U.S. citizens, and colonia populations are fairly stable.[4] Colonias are not an immigration problem; they are a highly inadequate solution to a shortage of affordable housing in an impoverished region.

Over time, the lack of subdivision regulations in Texas border counties has attracted Texas developers. According to Holz and Davies, colonias have developed abundantly in Texas because "little risk and high profits" are associated with their development.[5] As Holz and Davies explain, rural land has traditionally been used for agriculture: the associated risks and high costs have included preparation, maintenance, harvesting, and marketing. Given the cost and risk associated with farming, and the more reliable return from converting farms to homes, many border landowners have opted to subdivide and sell their land. Holz and Davies maintain that colonia developers enjoy low initial costs because these subdivisions lack traditional infrastructure, such as water and wastewater systems, generally associated with residential subdivisions. Initial development costs can be repaid by collecting downpayments on subdivided lots. Developers often sell colonia lots through contracts-for-deed, under which residents have no deed or claim on the property until they have paid for their lot in full. After the break-even point, developers can earn $7,200 per year through contracts-for-deed on an acre of land.[6]

Some developers may promise to provide water and wastewater infrastructure either verbally or in a contract at the time of purchase, but this promise is seldom fulfilled. In the rare instance that the promise is a part of a contract and legal action is taken on behalf of residents, developers have filed bankruptcy to avoid paying for infrastructure. Responsibility for providing services to colonias then falls to local and state governments. Davies and Holz quote an *Austin American Statesman* article in which a colonia developer states that "ensuring proper sewage is the job of the health department, not the developer."[7] The question of who is responsible for serving colonias is one of the key issues to be resolved by the state as its approach to colonia issues continues to evolve.

Texas has attempted to address the proliferation of colonias along the border by increasing control over the procedures that developers use to create these communities. The 1989 Model Subdivision Rules (MSR) were intended to arrest the development of new colonias by requiring that adequate water and wastewater facilities be provided in residential developments with plots of land of one acre of less. Developers also are required by MSR to submit a plat to the county, and they are restricted in the number of dwellings that can be built on each lot. The TWDB is prohibited from providing assistance under the Economically Distressed Areas Program (EDAP) to counties that have not adopted these rules, and it can suspend funding in counties that exhibit patterns of violation or other problems with compliance.

Part of the reason for the proliferation of Texas colonias is the relatively weak county government system.[8] Because colonias lack local governments, residents are governed by the counties in which they reside. Counties in Texas historically have not had significant authority to regulate development.[9] As a result, the state has had to finance the development of colonia water and wastewater service. Texas counties have long sought increased power to regulate colonia developers.

House Bill 1001, also known as the Colonias Bill, was passed during the 1995 Texas state legislative session. It significantly changed the regulatory and zoning authority of county governments. The bill authorizes counties to provide utility service, including water and sewer service, to colonias. House Bill 1001 also addresses another significant concern regarding environmental infrastructure for colonias. When the initial colonias laws creating EDAP and the Model Subdivision Rules were enacted in 1989, subdivisions that had been legally platted but not developed prior to July 1989 were exempted from the MSR. The requirements of HB 1001, however, apply to many of the platted subdivisions grandfathered under previous legislation. The removal of special status for many thousands of unsold lots is intended to prevent further growth in areas that lack adequate water and sewer infrastructure.[10] Another key issue to be resolved by the state involves the question of colonia legislation. Texas will need to determine whether existing state regulations are sufficient to halt the growth of colonias, and whether those regulations are enforced adequately.

Colonia Infrastructure Funding

Texas border colonias can access a variety of federal and state sources for subsidized water and wastewater infrastructure funding. Killgore and Eaton report state and federal estimates of existing funds for environmental infrastructure in the U.S.-Mexican border region that amount to approximately $2 billion; about $1.2 billion of that figure is accounted for by Texas Water Development Board loan and grant funds.[11] The TWDB, Texas Department of Housing and Community Affairs (TDHCA), Rural Economic and Community Development Agency (RECDA, formerly the Farmers' Home Administration and the Rural Development Administration), U.S. Department of Commerce, and U.S. Environmental Protection Agency (EPA) all provide financial assistance to colonia water and wastewater projects.

Many of these agencies also provide technical assistance and support through the programs described in this report and elsewhere. In addition, other federal and state agencies such as the Texas Department of Health and U.S. Department of the Interior may provide technical assistance and support to colonias, although they do not provide direct water and wastewater project funding. The agencies and programs described here are only those that support some form of direct project finance. This focus on infrastructure support rather than technical assistance is not intended to minimize the importance of technical assistance, however.

Listing programs under specific agencies can convey an incomplete picture of the interagency nature of many colonias initiatives. Some of the funding provided by the two state agencies most involved in colonia water and wastewater project finance, TWDB and TDHCA, is actually federal money. These agencies administer federal funds for colonias projects from EPA and the U.S. Department of Housing and Urban Development (HUD), respectively. Some smaller programs depend upon cooperation among a variety of public and private institutions. The Texas Small Towns Environmental Program (STEP) is coordinated through the Texas Natural Resource Conservation Commission (TNRCC). But STEP is funded by a combination of sources, including the TWDB (with interest earned from EPA funds), private organizations like the Pew Charitable Trusts and the Rensselaerville Institute, local governments, and colonia residents. STEP projects could not be proposed, developed, financed, constructed, operated, and maintained without the cooperation of these myriad institutions and communities.

These examples illustrate the difficulty of describing precisely the structure and quantity of international, federal, state, and local resources available for colonia water and wastewater infrastructure projects. Nonetheless, it is useful to examine the current status of major funding sources. Programs administered by TWDB and TDHCA are described first and in the greatest detail. The majority of colonia infrastructure funding comes through these two agencies. One TWDB program in particular, the Economically Distressed Areas Program (EDAP), is evaluated at length in order to illustrate its true impact on Texas colonias.

Texas Water Development Board Programs

Economically Distressed Areas Program (EDAP)

The main source of funding for colonia water and wastewater systems is EDAP, established in 1989 by the 71st Texas Legislature. EDAP serves Texas counties with a per capita income 25 percent below the state average and unemployment 25 percent above the state average for the last three years. Counties adjacent to an international border are eligible regardless of other criteria.[12] EDAP-eligible counties must adopt MSR standards. As of the 1995 TWDB needs survey, 14 counties had adopted MSR. Table 1 describes the process that subdivisions must follow in order to obtain EDAP funds for colonia infrastructure delivery.

Table 1
Process to Obtain EDAP Funds for Colonia Infrastructure Delivery

Needs Survey	Facilities Plan	Plans and Specifications	Final Cost
TWDB conducts preliminary study of colonias in border area.	Each political subdivision completes plan to deliver needed systems, including cost estimates.	Political subdivision commissions detailed survey of area with actual bids from contractors.	TWDB allots 115 percent of lowest construction design estimate.

Source: Unpublished interviews with EDAP staff by policy research project class members

Only political subdivisions can apply directly for EDAP funds. As most colonias are unincorporated, they need sponsors to access EDAP funds. A colonia or sponsoring political subdivision in an EDAP-eligible county must navigate a lengthy application process to obtain EDAP funds. The subdivision must complete a facilities plan, a detailed plan signed and sealed by a state-licensed engineer which is also considered a cost estimate.[13] The TWDB provides funding for this stage but does not consider most projects for eligibility until the completion of the facilities plan. Once a facilities plan is approved, the applicant is funded to perform a plans and specification study, which details the construction design and gathers bids for construction. Although the applicant does not specify a total cost estimate for the project in the plans and specifications study, the program is designed to award contracts to the lowest bidder. The TWDB uses these bids to fund projects, allotting 115 percent of the lowest bid estimate for the total cost of the project to allow for unforeseen costs.[14]

Table 2
EDAP Projects by County, 1995

County	Complete	Started	Planned*	Total
	EDAP Projects			
Bee	0	0	0	0
Brooks	0	0	0	0
Cameron	2	0	6	8
Dimmit	0	0	1	1
Duval	0	0	0	0
Edwards	0	0	0	0
El Paso	2	0	8	10
Frio	0	0	0	0
Hidalgo	1	2	18	21
Hudspeth	0	0	1	1
Jeff Davis	0	0	0	0
Jim Hogg	0	0	0	0
Jim Wells	0	0	0	0
Kinney	0	0	1	1
La Salle	0	0	1	1
Maverick	0	1	1	2
Pecos	0	0	0	0
Presidio	0	0	1	1
Reeves	0	0	0	0
San Patricio	0	0	2	2
Starr	0	0	2	2
Terrell	0	0	1	1
Uvalde	0	0	1	1
Val Verde	0	1	1	2
Webb	0	0	3	3
Willacy	1	0	1	2
Zapata	0	0	1	1
Zavala	0	0	2	2
Total	**6**	**4**	**52**	**62**

Source: Texas Water Development Board, "Economically Distressed Areas Program, Summary of EDAP Projects," Austin, Texas, February 15, 1996.

*Includes projects in the plans and specifications and facilities planning stages.

Table 3
EDAP Funds Distributed by County, 1995

County	Population Served by EDAP Projects*	Estimated Total EDAP Funding*	EDAP Funding per Capita
Bee	0	$0	$0
Brooks	0	$0	$0
Cameron	24,262	$52,336,800	$2,157
Dimmit	1,344	$2,837,254	$2,111
Duval	0	$0	$0
Edwards	0	$0	$0
El Paso	42,254	$106,128,549	$2,512
Frio	0	$0	$0
Hidalgo	80,782	$127,748,727	$1,581
Hudspeth	887	$1,240,000	$1,398
Jeff Davis	0	$0	$0
Jim Hogg	0	$0	$0
Jim Wells	0	$0	$0
Kinney	81	$588,000	$7,259
La Salle	960	$1,154,400	$1,203
Maverick	7,998	$13,098,125	$1,638
Pecos	0	$0	$0
Presidio	59	$420,250	$7,123
Reeves	0	$0	$0
San Patricio	9,430	$25,832,844	$2,739
Starr	7,378	$4,372,500	$593
Terrell	1,000	$2,851,675	$2,852
Uvalde	2,124	$5,623,605	$2,648
Val Verde	1,450	$15,520,000	$10,703
Webb	4,175	$11,005,066	$2,636
Willacy	2,728	$4,831,000	$1,771
Zapata	2,242	$5,051,000	$2,253
Zavala	4,404	$3,839,611	$872
Total**	**193,558**	**$384,479,406**	**$1,986**

Source: Texas Water Development Board, "Economically Distressed Areas Program, Summary of EDAP Projects," Austin, Texas, February 15, 1996.

*Includes completed, started, and planned projects.

**EDAP Funding Per Capita total is the state average per capita funding for EDAP projects.

Colonia Plumbing Loan Program

The Colonia Plumbing Loan Program (CPLP) allocates money to political subdivisions that, in turn, loan the money to individual community residents for laterals, hook-up fees, indoor plumbing, and the addition of bathrooms—"end-of-the-line" costs that are not funded by EDAP. Political subdivisions may borrow up to $4,000 times the number of qualifying households.[15] In addition, CPLP provides low-interest loans and up to 9 percent grants for administrative expenses within the subdivision, which must loan out the money and make reasonable efforts to collect payments. CPLP funds for border projects amounted to approximately $10.6 million as of February 1996.[16]

Texas Department of Housing and Community Affairs Programs

The Texas Department of Housing and Community Affairs (TDHCA) identifies the housing needs of colonia residents as "the most critical in the state of Texas."[17] TDHCA administers all funds allotted to Texas under the U.S. Department of Housing and Urban Development's (HUD) Community Development Block Grant (CDBG) program. Since the passage of the Cranston-Gonzales National Affordable Housing Act in 1990 (Public Law 101-625), HUD has required that Texas set aside 10 percent of its CDBG allocation for colonias. This 10 percent, amounting to approximately $9 million in 1996, is administered through TDHCA's Colonia Fund. Table 4 describes the disbursement of CDBG funds to colonias through TDHCA prior to June 1996, when the Office of Colonia Initiatives was created to manage all TDHCA colonias programs through TDHCA prior to June 1996, when the Office of Colonia Initiatives was created to manage all TDHCA colonias programs.

Table 4
Texas CDBG Funds Disbursement to Colonias

Federal Government (Department of Housing and Urban Affairs)
⇓
Texas Department of Housing and Community Affairs
⇓
Texas Community Development Program
⇓
Colonias Fund (10 percent minimum)
⇓
Eligible Applicants

Source: Unpublished interviews with EDAP staff by policy research project class members

Eligible counties may submit applications on behalf of a colonia located within 150 miles of the Texas-Mexico border. Applicants must submit the following in order for an application to be considered complete: (1) community needs assessment project summary; (2) number of low- and moderate-income persons project will benefit; (3) budget; (4) citizen participation plan; (5) public hearing information; and (6) project maps. Current and completed projects in the CDBG Colonia Fund database will affect 34 counties, listed in Table 5. Many counties that receive TDHCA colonia funds are also EDAP-eligible.

Table 5
Counties Involved in TDHCA Colonia Projects

Aransas	Hidalgo	Reeves
Bee	Hudspeth	San Patricio
Brooks	Jim Wells	Starr
Cameron	Karnes	Terrell
Crockett	Kleberg	Tom Green
Dimmit	La Salle	Uvalde
Duval	Live Oak	Webb
Ector	Maverick	Willacy
El Paso	Nueces	Zapata
Frio	Pecos	Zavala
Gillespie	Presidio	
Glasscock	Real	

Source: Texas Department of Housing and Community Affairs, Colonia Fund database, Austin, Texas, undated

As of April 1996, TDHCA had given grants for 96 projects, affecting 146 Texas colonias. Some affected colonias may be double-counted, because planning and construction grants are given separately, and a colonia is counted as "affected" each time a TDHCA grant or loan funds related projects. Table 6 illustrates total CDBG grant funds used by colonias since 1991 and the combined match from counties that apply for CDBG funds. Table 6 also shows the total number of beneficiaries and number of beneficiaries in low- and moderate-income brackets. Part of the reason for the discrepancy between total actual beneficiaries and proposed beneficiaries is that many projects have not yet been completed. Thus far, nearly 80 percent of set-aside CDBG funds have benefited low- and moderate-income colonia residents.

Table 6
TDHCA Colonia Funds Distributed, All Counties, January 1991-April 1996

CDBG Colonia Fund grant total	$ 32,163,201
Match from counties	$ 11,458,947
Proposed low/moderate income beneficiaries	50,600
Proposed total beneficiaries	61,477
Actual low/moderate income beneficiaries	17,677
Actual total beneficiaries	22,251

Source: Texas Department of Housing and Community Affairs, Colonia Fund database, Austin, Texas, undated

CDBG Colonia Fund Structure

The Colonia Fund awards grant assistance bi-annually on a competitive basis to colonia residents in eligible counties. From 1991-1992, TDHCA allotted CDBG colonia funding in two rounds, Colonia Fund and Colonia Fund II. Projects with first priority, the most competitive applications, were financed from the Colonia Fund, and second priority projects were financed with remaining funds from Colonia Fund II.

Beginning in 1993, TDHCA created three separate funds within the Colonia Fund: the Colonia Construction Fund, Colonia Planning Fund, and Colonia Demonstration Fund. TDHCA allots funds under these first two categories for construction and planning of water, wastewater, and other colonia infrastructure needs. The third category, the Colonia Demonstration Fund, provides funding to meet community development needs and to "encourage the leveraging of other funds to provide basic infrastructure and other community needs through a comprehensive approach" that may include public facilities and infrastructure improvements.[18] Colonia Demonstration funds have more stringent eligibility requirements: the area must already be platted, and at least 75 percent of lots in the area must be owned and occupied by individuals other than the developer.[19]

Since 1993, the Colonia Construction Fund has financed the largest number of projects, about 40 percent. The Colonia Construction Fund has also been the largest in terms of dollars; approximately 51 percent of all TDHCA colonias funds have been distributed through the Colonia Construction Fund since 1991. Table 7 describes the portion of total TDHCA colonia projects and dollars associated with each fund. Figure 1 shows TDHCA colonia projects by fund type. The number above each bar denotes the percentage of total projects financed by each fund.

Table 7
Description of Project and Fund Distribution, TDHCA Colonias Programs, 1991-1996

TDHCA Fund	Number of Projects Financed	Percent of Total Projects Financed	Funds Allotted to Projects	Percent of Total Funds Allotted
Colonia Fund	20	20.8 %	$ 7,316,409	22.7 %
Colonia Fund II	10	10.4 %	2,997,889	9.3 %
Colonia Construction Fund	38	39.6 %	16,640,003	51.7 %
Colonia Demonstration Fund	4	4.2 %	4,000,000	12.4 %
Colonia Planning Fund	24	25.0 %	1,208,900	3.8 %
Total	**96**	**100.0 %**	**32,163,201**	**100.0 %**

Source: Texas Department of Housing and Community Affairs, Colonia Fund database, Austin, Texas, undated

As of November 1995, construction activities were complete for 98 percent of contracts funded in program year 1991. Colonia Fund contracts funded in 1992 were approximately 63 percent complete, and projects funded during the 1993 cycle were 9 percent complete. None of the 1994 program year projects were complete as of November 1995. Although Section 916 of the Cranston-Gonzales National Affordable Housing Act of 1990, which mandated a 10 percent set-aside for colonias from Texas' state CDBG allocation, expired in September 1994, TDHCA continues to provide funding for construction activities. In the 1995 program year, TDHCA allocated $5.5 million to the Colonia Construction Fund. It is important to note that funds are actually made available in the year following the cycle in which they are awarded. For example, 1994 program year projects began to receive funding in May 1995.

The overall total funds awarded to eligible counties between 1991 and 1996 is nearly $32.7 million. Total grant amounts vary from county to county. Grant totals to individual counties range from $20,000 (Hudspeth) to $2.86 million (Cameron). Colonia Fund projects funded through April 1996 are in the process of serving 146 Texas colonias. Appendix B lists affected colonias, including detailed county-level summaries which specify project costs, start and completion dates, beneficiaries, and distribution of activities. Including current allocations and TDHCA forecasts of total program beneficiaries, the Colonia Fund proposes to survey and respond to infrastructure and community needs for 59,813 colonia residents at a total cost of $44,255,582.

Figure 1
TDHCA Colonia Projects by Fund

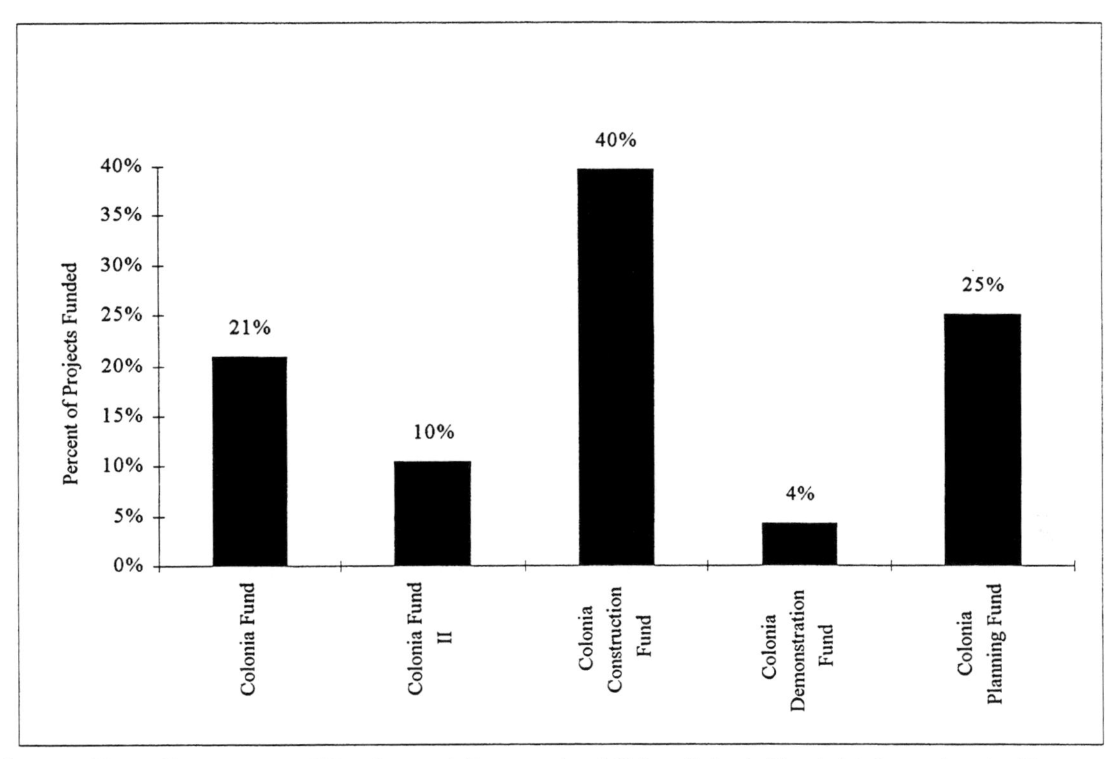

Source: Texas Department of Housing and Community Affairs, Colonia Fund database, Austin, Texas, undated

Activities Financed Through TDHCA Colonia Fund

The TDHCA pays for colonia projects related to the following services in eligible counties: water, sewer, septic tank, planning, street paving, drainage, community centers, housing, and comprehensive demonstration projects. Descriptions in the TDHCA colonia database show that these projects include rehabilitation, improvement, or first-time provision of the above activities. Projects which include access to water service and infrastructure are the most frequent, affecting 74 colonias. The second most frequent activity has been the provision of sewer service and infrastructure (not including septic tanks). Figure 2 shows the percentage of colonias affected by each type of TDHCA funding activity.

Table 8 describes TDHCA colonia infrastructure expenditures by county, including matching funds provided by each county for TDHCA projects and expected and actual project beneficiaries. For a county-by-county breakdown listing each colonia and its associated TDHCA project funds, please see Appendix B.

Figure 2
TDHCA Colonia Projects by Activity Type

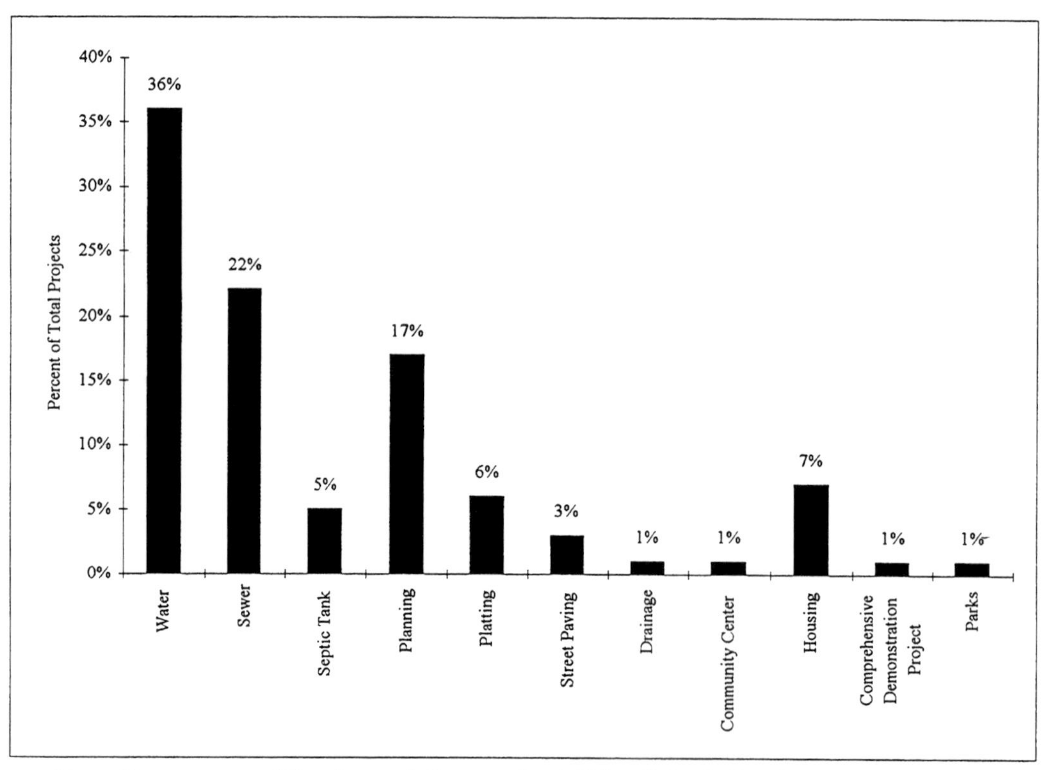

Source: Texas Department of Housing and Community Affairs, Colonia Fund database, Austin, Texas, undated

Other TDHCA Colonia Projects

In addition to the CDBG Colonia Fund, TDHCA administers the Colonias Initiative. The goal of the Initiative is to provide low-income home-ownership and affordable rental housing opportunities in colonias. Specifically, these programs strive to promote self-help techniques to "encourage resident participation, lower development costs and create new job skills."[20] TDHCA currently is developing the selection process for which and colonias will be included as Initiative Self-Help Center sites. These self-help centers will include community centers to provide health care, education, employment training, and counseling services. TDHCA has also allocated $2 million toward the TWDB's Economically Distressed Areas Program. These funds have been earmarked to help low- and moderate-income colonia residents pay for construction of water and sewer connections, yard lines, meters, and other costs associated with the provision of water and sewer service.

Table 8
TDHCA Colonia Funds by County, April 1996

County	Colonia Fund Grant Amount	Match from County	Proposed Number of Beneficiaries: Low/ Moderate Income	Proposed Number of Beneficiaries: Total	Actual Number of Beneficiaries: Low/ Moderate Income	Actual Number of Beneficiaries: Total
Aransas	$231,510	$27,000	92	93		
Bee	$499,845	$40,000	252	288		
Brooks	$1,410,700	$100,800	605	671	100	107
Cameron	$2,856,609	$93,901	10,294	12,329	3,557	4,595
Crockett	$495,800	$45,000	412	428		
Dimmit	$317,575	$15,675	187	211		
Duval	$297,890	$21,801	180	200	277	325
Ector	$27,500	$0	4,340	7,551		
El Paso	$2,000,000	$4,251,124	2,428	2,637	526	555
Frio	$636,500	$204,000	826	916	873	1,033
Gillespie	$300,000	$60,000	169	169	169	169
Glasscock	$542,250	$25,093	200	329		
Hidalgo	$2,338,538	$972,251	6,302	7,133	4,577	5,542
Hudspeth	$20,000	$1,208	71	73	71	73
Jim Wells	$530,750	$33,000	546	661	189	210
Karnes	$536,500	$0	574	618		
Kleberg	$766,920	$15,000	519	628		
La Salle	$506,500	$25,000	201	213		
Live Oak	$396,000	$20,000	141	145	160	211
Maverick	$2,325,981	$0	5,578	6,303	925	1,550
Nueces	$388,750	$195,000	1,709	2,364		
Pecos	$799,999	$41,973	383	454	226	254
Presidio	$1,800,000	$112,200	358	361	220	220
Real	$483,940	$25,844	237	270		
Reeves	$1,175,000	$55,000	916	1,345	555	555
San Patricio	$1,611,800	$115,000	2,083	2,104	1,530	1,574
Starr	$1,281,915	$0	795	839	63	63
Terrell	$56,750	$0	613	1,136	613	1,136
Tom Green	$425,881	$710,000	292	522		
Uvalde	$498,505	$2,995	251	282		
Webb	$2,186,070	$2,270,378	1,974	2,321	820	976
Willacy	$1,263,822	$1,824,704	2,338	2,620	1,213	1,213
Zapata	$1,300,000	$90,000	880	907	629	656
Zavala	$1,853,401	$65,000	3,854	4,356	1,731	2,097
Total	**$32,163,201**	**$11,458,947**	**50,600**	**61,477**	**17,677**	**22,251**

Source: Texas Department of Housing and Community Affairs, Colonia Fund database, Austin, Texas, undated

Texas Natural Resource Conservation Commission Programs

Although staff in many departments and regional offices of the Texas Natural Resource Conservation Commission (TNRCC) provide research, technical assistance, and support for colonia environmental infrastructure development, only two programs are involved in direct project financing. TNRCC is the state coordinating agency for the Texas Small Towns Environmental Program (STEP), which helps colonias and other Texas communities develop affordable, small-scale self-help projects. TNRCC cooperates with the Texas Department of Health, TDHCA, TWDB, and General Land Office with support and guidance from the national STEP program based in the Rensselaerville Institute. Water and wastewater projects are developed by residents in small communities by first determining what they can afford, and then mobilizing their own resources to accomplish the project. Additional project funds come from a variety of sources, including a $2 million set-aside at the TWDB (comprising interest earned on other funds and a matching grant from the Pew Charitable Trusts), project grants and loans from Waterworks (a consortium of charitable trusts), and the Rensselaerville Institute.[21]

The TNRCC also administers a program of Circuit Rider Technical Assistance for Public Water Systems along the Texas-Mexico border. Using two-year EPA discretionary funds and agency matching funds, this program provides concentrated technical and management assistance to public water systems in Texas counties along the Rio Grande. Objectives are to decrease the number of systems in substantial noncompliance with public drinking water standards and to ensure that systems provide reliable and adequate customer services. The TNRCC contracts with the Texas Rural Water Association to provide a trained circuit rider to visit all border public water systems at least twice over a predetermined contract period and to be on call to provide assistance as needed. The circuit rider also coordinates educational seminars.

Rural Economic and Community Development Agency Programs, U.S. Department of Agriculture

The Rural Economic and Community Development Agency (RECDA) administers funds for colonia water and wastewater infrastructure development under two separate programs: Water and Wastewater Grants and Loans, and Section 306(c). Under the first program, RECDA funds installation, repair, or improvement to water and sewer systems (including laterals and connection assessments) for rural communities, defined as areas with a population of 10,000 or less. The program provides combinations of grants and loans; grants are awarded up to a maximum of 75 percent of eligible project costs and only when necessary to reduce average annual user fees to a reasonable level. During the 1996 fiscal year, Texas was allocated $22.8 million in loan funds and $13.5 million in grant funds under this program.[22] Of that amount, approximately $4 million in loans and $3 million in grants were expected to be used for border water and wastewater projects.

RECDA's Section 306(c) program funds similar projects, but eligible communities are restricted to colonias with populations under 10,000. Grant funding is available up to 100 percent of project costs, but funds are limited. RECDA stretches Section 306(c) funds by combining them with funds from other programs. Texas colonias received an FY 1996 Section 306(c) appropriation of $6.3 million.[23] RECDA also has some housing programs that may provide funding for bathroom improvements and general home improvements. These programs provide direct loans and grants from the agency to individuals.

Economic Development Administration Programs, U.S. Department of Commerce

The Economic Development Administration (EDA) funds colonia water and wastewater improvements through two separate programs: Economic Development Grants and the Public Works Impact Program. Economic Development Grants fund public facilities, including water and wastewater systems. Public works must be tied to the creation or retention of permanent private sector jobs. Local matching funds are required; the average rate is 50 percent grant funding. Severely depressed areas may receive up to 80 percent grant funding. Individual project financing usually ranges from $80,000 to $2 million.[24]

The Public Works Impact Program funds renovation or construction of public works to assist in providing immediate useful work to unemployed and under-employed persons in project areas. Eligible cities or counties must be designated as redevelopment areas under the Public Works and Economic Development Act. This program has matching requirements similar to the Economic Development Grants described above. Priority is given to projects of $600,000 or less.[25]

Both of these EDA grant programs are funded through a budget allocation of $30 million for a five-state area including Louisiana, Arkansas, Oklahoma, New Mexico, and Texas. The Department of Commerce assumes that Texas projects receive at least one-fifth of the total, or $6 million, because of the state's large population and the economic distress common to the border region.[26]

Environmental Protection Agency Programs

Since 1993, the Colonia Wastewater Treatment Assistance Program (CWTAP) has provided grants, through the TWDB, to local governments and nonprofit water supply corporations for design and construction of wastewater collection and treatment facilities.

In addition to the funds that EPA contributes to TWDB and TNRCC programs, this federal agency also administers directly at least two additional programs that contribute to water and wastewater infrastructure finance for Texas colonias. The Water Pollution Control, Research, Development, and Demonstration program supports projects relating to water pollution. Grant funds cannot exceed 75 percent of project costs. Individual project costs range from $9,000 to $580,000, and average just over $110,000.

The Safe Drinking Water Research and Demonstration program supports (among other goals) the provision of dependable, safe drinking water supplies, and prevention of disease due to contaminated water. Funds can be used to finance demonstration projects through grants, but the program requires local matching funds. Individual projects range from $3,000 to $429,000 and average approximately $78,000.

NAFTA-Related Programs

Following negotiations on the North American Free Trade Agreement (NAFTA), the United States and Mexico created a pair of institutions, the Border Environment Cooperation Commission (BECC) and the North American Development Bank (NADBank), designed to respond to border community environmental infrastructure needs. The two are designed to work together; BECC solicits project proposals and channels appropriate projects to NADBank. The Bank develops public and private sources of financing for the projects, injecting some of its own funds when projects fit its criteria.

It remains to be seen whether BECC and NADBank will be significant resources for Texas colonias. Through November 1996, BECC had received over 90 project proposals, considered 65 that met its basic criteria, and certified 12 projects.[27] Through December 1996, NADBank had approved financing packages for four projects (two U.S. and two Mexican), injecting $31 million of its own funds. Three Texas projects have been approved by BECC, and one has received NADBank funding.[28]

Two of the Texas projects approved by BECC will have direct impact on colonias in El Paso County. The first, a $11.7 million wastewater reuse project, will recycle treated wastewater for irrigation and industrial uses in Northwest El Paso, benefiting 90,000 residents. The second Texas project is a $110,000 on-site, self-help wastewater treatment system for El Paso colonias. The project will provide no-interest loans to help 250 colonia families properly install septic tanks and treat household sewage.[29]

BECC and NADBank have not bridged the gulf between the need to create a viable international financial institution with access to international bond markets, and the need to create a funding source accessible to low-income border residents. NADBank money is expensive relative to financing by TWDB, and some analysts believe that NADBank's repayment criteria are unrealistic.[30] The Bank's main strength is that it may lend to both public and private entities with sound proposals. Private institutions cannot access the subsidized assistance provided through the federal and state programs described above. In addition, Texas colonias may benefit from NADBank funding when projects are unable to obtain funding from another source; when private firms find NADBank interest rates competitive; when proposed projects do not meet the eligibility requirements of state and federal programs; when projects funded by other public and private sources might use supplementary NADBank funding or loan guarantees; or when necessary components of a project cannot be covered by most existing sources.

This is not a comprehensive list of sources of funding for water and wastewater infrastructure projects in Texas border colonias. Many communities, for example, apply to private entities like the Ford, Kellogg, and Mott foundations for infrastructure improvements. But these are the largest of the state, federal, and international programs that provide funding for Texas border water and wastewater projects. It is beyond the scope of this report to calculate the exact amount of funding available under these programs.

Remaining Needs

The work of these myriad state, federal, and international colonia infrastructure finance sources will not ensure adequate service. When all funds appropriated for colonias programs have been exhausted, three categories of remaining needs will continue to exist: (1) colonias ineligible for funding from the most substantial existing programs (for example, those in counties which have not adopted the Model Subdivision Rules); (2) colonias that are eligible, but to which extension of water and wastewater service is not cost-effective for public or private sector institutions; and (3) colonias that are both eligible and cost-effective to serve, but that for some other reason have not yet been reached. The State of Texas may face even greater challenges trying to meet the needs of remaining colonias than it has in designing current colonia infrastructure programs.

The TWDB is the only institution that has calculated and published an estimate of aggregate colonia water and wastewater infrastructure needs that remain to be addressed. The Board's 1992 needs survey estimated a total cost of $696 million to provide water and wastewater services to colonia residents, including indoor plumbing and connection costs. The TWDB's *Water and Wastewater Needs of Texas Colonias, 1995 Update* estimated a remaining need of $424.6 million, not including indoor plumbing and connection costs. Although these estimates represent a remarkable contribution by the TWDB, they remain rife with problems, indicative of the difficulty of describing colonia needs.

TWDB Needs Surveys

In 1992, the TWDB completed the first aggregate needs survey of Texas colonias, *Water and Wastewater Needs of Colonias in Texas* (1992), in which ten TWDB employees conducted field visits over 46,000 square miles in nine months. This process was complicated by a number of obstacles. First, the nature of colonias makes needs measurement extremely difficult. Colonias are hard to locate and are spread throughout the Texas-Mexico border region, an area the size of the state of Louisiana. The EDAP staff is small, as is the TWDB, and the TWDB did not have the resources to make more accurate assessment feasible. Limits of the number of personnel and time prevented precise measurement of needs, contributing to a number of methodological weaknesses.

The TWDB's 1992 and 1995 needs surveys attempted to estimate the total amount of capital expenditure required to meet colonia needs. The board approached the project as the first general study of Texas colonias, and the 1992 survey cautions that the study was "limited in scope."[31] In fact, both studies were conducted on a "drive-by" basis, requiring the use of individual judgment and estimation.[32] These TWDB surveys are, however, the most comprehensive state estimates of the aggregate costs of delivering water and wastewater infrastructure to Texas colonias.

In the 1992 survey, the TWDB estimated total costs beginning with individual or groups of colonias. Estimates were made for each colonia based on the following five factors:

- condition and adequacy of current infrastructure;
- either the distance between a water supplier and the colonia, or the distance between the central point of a group of colonias and each colonia in the group;
- either the distance between a wastewater treatment facility and the colonia or group of colonias, or potential for an alternative wastewater treatment/disposal method;
- need for indoor plumbing or plumbing improvements; and
- costs for connection to services.

The TWDB used average costs throughout the survey, including the cost per foot of pipeline and cost estimates for alternative wastewater treatment systems, plumbing installation, home-to-curb hookup, and home-to-meter hookup.[33]

TWDB staff relied heavily on the individual engineering judgment of ten field workers to establish populations, housing numbers, lot sizes, feasibility of technologies, and adequacy of existing infrastructure. Relying on personal judgment inevitably leads to inconsistency; different people make different estimates, especially for population, number of houses, lot size, and distance.[34] In an attempt to lessen the influence of personal judgment, however, the TWDB summarized these estimates and sent them to county officials for verification. Only after figures were reviewed by local authorities did the TWDB proceed with cost calculations.[35] In addition, for the 1995 update, many colonias in the 23 counties resurveyed were already in the facilities planning stage.[36] In many cases, local engineering or consulting firms had completed detailed analyses of colonia characteristics using local knowledge and information, which provided more accurate data for the 1995 update. The TWDB gathered other data from the EPA's Colonia Wastewater Treatment Assistance Program (CWTAP) and TDHCA.[37]

Central Water Systems

Estimating the cost of a pipeline for central water systems is one of the complications of calculating colonia infrastructure needs. The Board cannot simply measure the distance from a colonia to the nearest water or wastewater treatment facility and calculate the cost of installing a pipeline over that distance. Because water and wastewater facilities cannot be required to serve an unincorporated community, the closest facility will not necessarily fulfill the needs of a given colonia. Nevertheless, approximation of the distance from a colonia to an appropriate water treatment facility is a crucial step in estimating the cost of connecting a colonia to a central system. After approximating distance, the TWDB applies concrete rules to cost estimation. For the 1992 study, the TWDB used the following figures to estimate the cost of water delivery systems:

- $18 per foot for transmission lines from water supplier to colonia; and
- $12 per foot for distribution lines within colonias.[38]

Applied uniformly, these figures provide a reasonable guess at the costs of ensuring safe drinking water for every resident of a colonia. However, it is difficult to determine the distance from a water supplier to a colonia when it is unclear which water supplier will serve that colonia. Even if the board were to work from the unrealistic assumption that colonias will receive water from the closest supplier, TWDB staff did not have the time or the resources in these two surveys to determine the exact distance to be covered by transmission lines and relied heavily on estimates made by the ten field workers both on-site and in the office using maps. The impact of errors on the total 1992 needs estimate could, therefore, be substantial.

Central Wastewater Systems

For wastewater treatment, TWDB staff compared the feasibility and cost of a range of potential solutions for colonias. According to the 1992 survey, "for larger, more densely populated settlements, regional and/or centralized treatment systems were considered viable options."[39] Colonias were grouped by the TWDB staff to estimate wastewater needs in order to consider this option. Colonias which were grouped could be expected to connect to the same wastewater treatment plant. The 1992 survey further explains that regional treatment would come from an existing wastewater treatment plant, and centralized treatment would require construction of a new plant, centrally located to serve all colonias in the group. Implicitly, the TWDB assumed that clusters of colonias will jointly build or use a wastewater treatment plant; that may or may not occur. It is difficult to encourage political subdivisions to apply for EDAP funds to assist colonias. Once they do, there is no guarantee that they will apply to serve the same colonias that the TWDB grouped into one survey area. The TWDB estimated traditional system costs at:

- $26 per foot for the interceptor line;
- $23 per foot for the collection line within the colonia; and
- $35 per unit per year for oversight and management of collection lines.[40]

Wastewater collection system cost estimates are subject to the same distance estimation problems that occurred for water delivery systems. That is, the nearest or least costly wastewater treatment facility may not, in the end, be the facility to which a colonia is connected.

Other Wastewater Treatment Systems

For smaller and more rural colonias, TWDB staff considered on-site wastewater treatment, such as septic tanks or small cluster systems, and estimated costs for these systems at:

- $640 per capita for a septic tank and drainfield and $35 per unit per year for system oversight and management; or

- $6,000 per unit for an evapotranspiration system with $35 per unit per year for system oversight and management.[41]

The TWDB study assigned systems to colonias based on general reviews of location, soil conditions, and availability of nearby services. The study cautions that estimates are preliminary.[42] Any construction done to meet colonia needs would have to be based on a detailed engineering study.

Plumbing and Service Connections

The final costs included in the 1992 needs survey were the costs of connecting residents to water delivery and wastewater removal systems, as well as providing indoor plumbing where necessary. These costs are not included in the 1995 needs estimate because the funding and administration of these costs had been moved from EDAP to the Colonia Plumbing Loan Program, also administered by the TWDB.

Needs Survey Weaknesses

Due to time and personnel constraints, the TWDB was not able to measure exact distances from colonias to water suppliers or wastewater treatment facilities, nor were they able to measure exact sizes of colonia lots, which can determine eligibility for alternative wastewater treatment methods. TWDB staff estimated distances through map inspections and "individual engineering judgment."[43] Based on TWDB methodology, accurate cost estimation for colonia water and wastewater systems requires data including: (1) condition of current infrastructure; (2) colonia population; (3) colonia location; (4) who will supply water or who will collect and treat wastewater; (5) pipeline and installation costs; (6) current indoor plumbing status of colonia homes; and (7) residents' ability to pay for home-to-curb and home-to-meter hookups. Measuring each of these factors is time consuming, expensive, and imprecise.

Accurate identification of current infrastructure in each colonia was beyond the scope of the two TWDB surveys. In the unlikely event that the TWDB could locate each individual colonia, traveling to all colonias, spread over 46,000 square miles, would have been impractical.[44] Future surveys will likely make use of information available from local utilities, water supply corporations, local governments, councils of governments, taxing districts and community-based organizations, as the TWDB has done. Developing a needs estimate on the basis of current infrastructure is also problematic. Current infrastructure may meet minimal state standards, but it may not be sufficient for water supply corporations, cities or utilities to undertake infrastructure expansion or modification. Most companies will include the cost of replacing, or substantially improving, the current delivery system in facility plans, resulting in higher actual costs than projected by the needs survey.

Total colonia population estimates range from 70,000 to 360,000 residents. Many of these thousands of colonia residents rely on migrant farm work for subsistence. When residents leave to work in other parts of the country, population estimates may drop drastically. Local, regional and community-based organizations can be helpful in gathering population data, although piecemeal gathering of data may lead to inconsistent methodology and unreliable results. Census data, although extremely detailed and helpful, may contain gaps in information due to lack of mail service in some colonias.

Timely collection of precise information on individual communities is made difficult by the large number of existing colonias, at least 1,436 in the border area, and their often remote location. Colonias vary in size and grow quickly. Determining location and the water supply corporation or utility which owns the water rights for an individual colonia is not impossible, but many water supply corporations and utilities are not keen to develop and implement colonia services. Some municipalities or other political subdivisions will find it advantageous to serve colonias due to the incentives provided by EDAP, but these communities may find it difficult to conclude agreements over sales of water rights. The costs of these rights should be included in any colonia infrastructure cost, but due to the difficulty of determining which colonias will be served by their current water rights owner and which rights will need to be negotiated, water rights costs cannot usually be included in cost estimates.

The TWDB usually incorporates average costs for pipeline and installation into its estimates. Problems arise, however, because variations in labor costs can occur throughout the border region and current costs will not necessarily equal future costs. The problems of inflation and rising labor costs could be avoided by adjusting to a constant dollar when calculating cost estimates.

The current indoor-plumbing status of dwellings in colonias is also difficult to determine. Colonia homes often are built piecemeal over long periods of time. Additions are constructed when residents have saved enough money to build them. Families surveyed during the data gathering stage may have almost enough money to install indoor plumbing at the time of the survey, but may be unprepared to do so until they have all of the money. Data about these homes will not include impending or potential projects. It is difficult to establish families' ability to pay for water and sewer hookups. Refusing to borrow money for hookups is extremely common in the border region, so needs estimates must set aside grant funds for these small-scale, household-level projects.

In spite of all of these weaknesses, the TWDB did complete the first aggregate study of colonia water and wastewater infrastructure needs. Through the board's effort, this report has a basis for comparing TWDB cost estimates to other estimates and to actual costs. One other estimate is provided by projects that have already completed facilities plans.

Facilities Plans

Facilities plans are generated for a colonia or a group of colonias when a political subdivision applies for EDAP funds. Facilities plans are detailed engineering studies of the colonia(s) and sponsor of the project. A general overview of the facilities plan methodology can give a broader understanding of the differences in cost estimates between the needs survey and the facilities plan. According to Steve Mendoza, EDAP

Engineering Specialist, infrastructure costs in a facilities plan are based on population. Because the plan is completed locally, by people intimately familiar with the area, population figures in facilities plans are believed to be reliable.[45]

For the facilities plan, a physical survey of the area is completed by an engineer able to measure exact distances from the appropriate water or wastewater facility to the colonia. Engineers are also able to measure internal distribution and collection distances. In addition, the plan requires knowledge of the number of lots in the colonia, existing population, average number of persons per dwelling, number of unoccupied houses, number of vacant lots, and growth rate for the area. The growth rate is generally the Census Bureau growth rate for the area or the growth rate as reported by The University of Texas, Pan-American. The facilities plan then computes two expected populations from this information: one extends the existing population on the basis of number of houses and growth rate for 20 years; the second calculates the buildout based on the number of unoccupied lots.[46]

The facilities plan model is a demand model, basing water delivery needs on expected community usage, through the smaller of the two population forecasts described above at 120 gallons per person per day. Wastewater removal is estimated at 75 to 85 percent of water demand. The model design and cost at this point are based on the engineer's judgment and experience. Facilities plans account for various wastewater technologies by determining the location, type and parameters of nearby streams, but almost consistently decide on traditional treatment.[47] The construction of treatment methods is estimated to cost about $1.00 per gallon capacity for pond treatments with no mechanical infrastructure, $2.00 per gallon capacity for activated sludge treatment and $2.50 per gallon capacity for extended activation.[48]

The facilities plan's population figures are extremely important in determining the total model cost. If current population or expected growth is determined inaccurately, the total expected cost will be inaccurate and will differ from actual cost in proportion to the accuracy of population estimates. In addition, the use of 20-year population projection estimates can be problematic. This issue will be discussed in greater detail later in this report.

Needs Survey and Facilities Plan Differences

The difficulty of comparing a facilities plan to a needs survey is threefold. First, when TWDB staff estimated colonia needs in the 1992 survey and 1995 update, any community already being served by a water delivery system which met state standards was not considered to be in need. Many colonias are rural settlements and had previously obtained water delivery through a Farmers' Home Administration program which implemented water delivery systems with two-inch diameter transmission lines to rural settlements. When TWDB survey staff visited a colonia receiving water through an FmHA-funded system, the colonia was not reported in need of water infrastructure.[49]

Second, when political subdivisions apply for EDAP funds, many are influenced by water supply corporations and utilities to include a new water delivery system to improve service, making it more profitable for the water supplier. A system that was determined to have met the needs of residents and was therefore not counted in the TWDB needs estimate may be discarded at the facilities plan stage in favor of a newer and better system with six-inch or larger diameter transmission lines. This "revisionist thinking" causes an increase in total costs from the TWDB needs survey to the individual facilities plan.[50]

Third, engineers who design facilities plans will be paid a percentage of total project. Engineers, therefore, have an incentive to keep project costs high. It is difficult to determine which facilities plans include new or improved water delivery systems in the best interest of the residents and which plans include new systems to inflate project costs. According to TWDB staff, the end result is that costs projected by facilities plans are often higher than those projected by the needs survey.[51]

The differences between the 1992 needs survey, facilities plans, and final costs for completed EDAP projects are quantified in Table 9. The projects listed in Table 9 do not demonstrate the discrepancy noted by TWDB staff. That is, the facilities plan cost estimates for these projects are not higher than the needs survey cost estimates. In all but one case, water infrastructure for Cameron Park, the facilities plan cost estimates are lower than the needs survey cost estimates. No conclusions should be drawn from this limited data, especially since most of the projects discussed had already been funded by the time TWDB estimated their total cost, but it is also possible that facilities plans understate actual cost estimates to ensure TWDB approval.

Comparing the needs survey and facilities plan cost estimates to actual project costs is also useful. Four out of five of the projects listed in Table 9 had already been funded through EDAP by the time the 1992 needs survey was completed, so cost estimates are closer to actual costs than would be expected of typical estimates. Two of the five needs estimates accurately predicted final expenditures, two underestimated final costs, and one overestimated final costs. Facilities plans for these projects underestimated both the 1992 needs survey and final costs by a substantial amount.

Whether they overestimate or underestimate actual project costs, facilities plans do not seem to provide, in these cases, information that can be aggregated to estimate total colonia infrastructure costs or other summary data. From a planning or policy perspective, neither the TWDB needs survey nor the individual facilities plans designed by engineering firms support the kind of informed decision-making that would be ideal for developing a state colonia infrastructure strategy. In the same vein, neither the needs survey nor the facilities plans provide an accurate platform from which to measure remaining colonia infrastructure needs.

Table 9
Sample Differences Between Needs Survey and Facilities Plans

	Cost Estimates		Final Cost	Cost Estimate Differences	
Colonia	1992 Needs Survey	Facilities Plan		Needs Survey-Final Cost	Facilities Plan-Final Cost
Cameron Park*					
Water	$2,215,000	$2,219,790	$2,856,076	($641,076)	($636,286)
Wastewater	$3,900,000	$3,246,192	$3,797,142	$102,858	($550,950)
Total	$6,115,000	$5,465,982	$6,653,218	($538,218)	($1,187,236)
Haciendas Gardens*					
Water	$0	$0	$0	$0	$0
Wastewater	$477,800	$434,683	$477,800	$0	($43,117)
Total	$477,800	$434,683	$477,800	$0	($43,117)
El Paso County Lower Valley WDA					
Water	$15,526,000	$11,203,900	$13,477,000	$2,049,000	($2,273,100)
Wastewater	$33,492,835	$28,829,200	$27,177,512	$6,315,323	$1,651,688
Total	$49,018,835	$40,033,100	$40,654,512	$8,364,323	($621,412)
Westway El Paso County*					
Water	$889,800	$667,185	$1,495,907	($606,107)	($828,722)
Wastewater	$0	$0	$0	$0	$0
Total	$889,800	$667,185	$1,495,907	($606,107)	($828,722)
Lull *					
Water	$590,000	$468,990	$590,000	$0	($121,010)
Wastewater	$860,000	$240,606	$860,000	$0	($619,394)
Total	$1,450,000	$709,596	$1,450,000	$0	($740,404)
Total	**$57,951,435**	**$47,310,546**	**$50,731,437**	**$7,219,998**	**($3,420,891)**

Source: Texas Water Development Board, *Water for Texas: Water and Wastewater Needs of the Colonias in Texas*, Austin, Texas, October 1992 and Facilities Plans submitted to the TWDB for Cameron Park, Haciendas Gardens, El Paso County Lower Valley WDA, Westway El Paso County, and Lull colonias.

*Cameron Park, Haciendas Gardens, Westway, and Lull had already been funded at the time of the 1992 needs survey.

Current Cost Estimates

The TWDB's 1992 needs survey estimated a total cost of $696 million to provide water and wastewater services to colonia residents, including indoor plumbing and connection costs. The 1995 update estimates a *remaining* need of $424.6 million, not including indoor plumbing and connection costs. The word *remaining* is emphasized to indicate that the cost of serving the colonias is not decreasing. Rather, it is still on the rise. However, the purpose of the TWDB needs survey was to estimate total costs needed to serve colonias not currently being served. Colonias included in completed facilities plans and those for which funding has already been set aside (those completed, under construction or included in engineering studies) have been removed from the 1995 estimate of costs to serve colonias. Thus the 1995 estimate of $424.6 million does not include $432.7 million which has already been committed or spent. The bottom line shows that the estimated costs of delivering basic water and wastewater infrastructure for the colonias rose by over $242 million between the 1992 and 1995 surveys. This does not include the funds taken out of EDAP to create CPLP, which would make this difference even greater. This may reflect growing colonia populations, outgrowth or development of new colonias, or identification of colonias previously not included. This could also be a result of the more precise distance measurements gleaned from facilities plans completed between 1992 and 1995, giving a more realistic needs estimate.

Regardless of the numerous problems associated with estimating colonia water and wastewater infrastructure needs and their associated costs, the well-being of Texans living in colonias depends in large part upon this process. The preceding analysis of the obstacles and challenges to determining needs should not be construed as a condemnation of the effort.

EDAP Status

On January 29, 1996, the TWDB testified to the Texas State Senate Committee on International Relations, Trade and Technology that almost 65 percent of colonias and more than 75 percent of colonia residents will be served by projects funded through EDAP. This information is also reported in the TWDB's *Water and Wastewater Needs of Texas Colonias: 1995 Update*. Neither the testimony nor the report specifies the varying stages of progress among colonias, but the report does state that this figure includes both "active projects" and projects in the facility planning stage of the EDAP application process. Although the report does not define an "active project," it differentiates between funded projects and projects in the planning stage. Thus, "active projects" can be interpreted to mean projects for which funding has been spent, committed by contract, or set aside for the sole purpose of implementing water and/or wastewater infrastructure in a colonia or colonias. Projects in the planning stage can be defined as those for which infrastructure improvement plans are supported by a political subdivision that has been funded through EDAP to complete a facilities plan.

Aggregate data for Texas colonias population does indicate that nearly 65 percent of the population of identified colonias will be served by an EDAP project and that these projects will include 75 percent of the colonias (the "65/75" impact). Estimates are divided in the report into the categories "funded" and "planning." The TWDB report includes the sum of "funded" and "planned" projects to calculate the 65/75 impact. According to the 1995 update, more than 100 percent of the population in Frio, Willacy, and Zavala counties and more than 100 percent of Frio county colonias will be served by EDAP projects. In addition, 200 percent of population and 200 percent of the colonias in Terrell County will be served by EDAP projects. These numbers suggest that some double-counting may have occurred in the 65/75 figure.

Due to the magnitude and breadth of the EDAP program, it is conceivable that these statistics, probably the result of double-counting, may have been overlooked. However, this contradiction at the county level reveals program evaluation problems at the colonia level. Colonias, as unincorporated subdivisions, lie within extra-territorial jurisdictions, like municipal utility districts. Yet colonias must be represented by a political subdivision applying for EDAP funds in order to be served by the program. Applying to serve a colonia with water or wastewater infrastructure may not be attractive to most political subdivisions because of the high poverty rate among colonia residents. Cities, utilities, water supply corporations, or other entities, often choose not to apply for EDAP funds because they are fearful that they will not be able to recoup the costs associated with providing service. Colonia residents' ability and willingness to pay for water and wastewater service has proven difficult to determine.[52] The limited financial resources of colonia residents, paired with the potential billing difficulties in a population perceived to be ambient are disincentives for service delivery by public and private institutions.

In order to make projects more financially attractive, applications for EDAP funds may include only water service or only wastewater service, even though most colonias need both services. The end result could be that two applications may be made by two different political subdivisions to serve the same colonia and same population. When this occurs, those served are counted twice by the TWDB as funded or planned in the 1995 survey update. It is possible that even three or more projects could potentially serve the same population and location. Thus, aggregate data indicating EDAP's impact on 65 percent of colonia residents and more than 75 percent of colonias may overstate the program's actual impact.

EDAP will *not* serve 75 percent of colonia residents and almost 65 percent of colonias when all projects, including those in the facilities planning stage, are completed. Rather, the program will only serve 59 percent of colonia residents and 56 percent of colonias in the Texas-Mexico border region. In seven years, EDAP has completed construction on only six projects, benefiting eleven colonias. More than 61 percent of projects, encompassing more than 76 percent of eligible colonias, are still in the facilities planning stage. When reporting the number of colonias and residents to be served, the TWDB did not account separately for colonias and residents served by more than one EDAP project.

A Better Estimate of Unmet Needs

The tables in Appendix A avoid the double- and even triple-counting described above by counting each colonia only once as "served," rather than counting it each time an EDAP project affects that colonia. Figures for previously-served colonias are not included in the totals for their other associated projects. In addition, each table lists all colonias identified by the TWDB needs database that are not affected by EDAP Projects. The tables contain a column entitled "number of people served by project." This column is used to calculate the total number of colonia residents who will benefit from the EDAP project. The number was calculated using the total number of people in need of the service proposed by the project. In some cases, both water and wastewater will be provided by the project. In these cases, if the entire colonia population is in need, the entire population will be served. If, however, not all residents are in need, the total number of residents in need of both services and the entire population are compared. The lesser of the two numbers is then used to indicate the number of people served. The assumption that residents in need of water infrastructure are not the same residents in need of wastewater infrastructure may overstate the total number of residents served. But this does not occur often and does not involve a significant number of residents, and therefore should not significantly impact totals.

The detailed information in Appendix A may be useful for some readers, but the most important points gleaned from the information are presented here. According to the tables presented in Appendix A, 356,164 residents live in 1,482 TWDB-documented colonias in 28 counties in the border area. Sixty-two EDAP Projects will serve 209,716 residents of TWDB-documented colonias, or 58.9 percent of all TWDB-documented colonia residents in the border area. This is 16 percentage points less than the figures the TWDB has published. In addition, EDAP projects will serve 831 colonias, or 56.1 percent of all colonias in the border area, or almost 9 percent less than the TWDB maintains the program will serve. The TWDB's publication, "Economically Distressed Areas Program, Summary of EDAP Projects" (as of February 15, 1996) states that 262,027 residents of colonias will be served by the projects tabulated in Appendix A. Subtracting 20,147 residents of "colonias" in East and North Texas County Projects, which are not tabulated in Appendix A, the TWDB figure states that 241,880 border residents will be served by EDAP projects.

Potential Explanations for Overestimation of Needs

The difference of 32,164 residents between the total from Appendix A and the TWDB figure can be attributed to many sources, including the problem of double counting. A second reason for overestimation is the 20-year population projection commonly used in the facilities planning stage.[53] The facility plan, as mentioned above, uses population projections for 20 years based on the area growth rate or the population of the colonia at buildout, whichever is less. Using a projected population in tabulating the number of people served does not, on the surface, present any problems. When calculating *percentages* of residents served, however, it does present a problem. If the total number of residents is not projected for 20 years or for complete buildout, any calculation using projected population may inflate estimates of the percentage of residents served. In other words, if populations are *projected* and counted as served, then the TWDB is including *future* colonia residents in its estimation of populations served. If the TWDB does not also include future residents, then, in the base number of colonia residents, percentages of residents served by EDAP will be upwardly biased.

Aside from the problem of inflated calculations, another factor could make the true statistics differ from the numbers presented in this report. It has been asserted that some subdivisions included in the TWDB needs database are not true colonias. Robert Goodwin and Ken Jones, property developers in Hidalgo County, are concerned that several, if not many, of the colonias listed in the needs database do not fulfill the basic definition of a colonia. Due to recent legislation which may burden them with the financial costs of constructing central wastewater systems, Goodwin and Jones commissioned a private review of all colonias in Hidalgo County that were included in the TWDB 1992 needs database. They assert that ten RV parks, three country clubs, one cemetery, five trailer and mobile home retirement areas, and one sports club are incorrectly listed in the TWDB database. In fact, a "colonia" that Goodwin and Jones insist is a cemetery, Highland Memorial Park, has been included in the Hidalgo County Urban Regional Planning facilities plan. If their assertion is true, striking non-colonias from the TWDB database could reduce the total number of colonias and increase the percentage of colonia residents served.

According to Augustine Tambe of the TWDB, however, engineers preparing facilities plans conduct field visits to all sites to determine eligibility for EDAP funds. Under these circumstances, it seems unlikely that a cemetery with no live residents would be classified as a colonia and included in a facilities plan. In addition, the TWDB verifies that all colonias included in a project do, in fact, meet EDAP requirements after the submission of the facilities plan and prior to approval of any financial commitment.

Appendix A also contains cost data. It will cost $384,479,406 to finance 60 projects serving 196,558 residents in 766 colonias. This results in averages of $6,252,267 per project, $489,734 per colonia, and $1,909 per resident. Using a density calculation of 4.399 persons per household, EDAP projects will cost an average of $8,396 per household. For these calculations, two projects were not included: Webb County (Southwest Webb) serving 61 new colonias and 7,288 residents and City of Mission (North Mission) in Hidalgo County serving four new colonias and 5,870 residents. These projects could not be included in cost calculations because cost estimates for project completion were not yet available.

Of the total number of current EDAP projects, 11 are complete, 38 are under construction, 147 are in the plans and specifications stage, and 635 are in the facilities planning stage. It is impossible to predict time ranges for the completion of projects, since each project has unique components. However, the plans and specifications stage has been reported to take anywhere from six to eighteen months. Due to the large concentration of EDAP projects in the facilities planning stage, this report asserts that most colonia residents may not actually receive water and wastewater service for some time.

Who is Not Yet Served?

Although many government, non-profit and private agencies are working to provide water and wastewater service to colonia residents in Texas, there is still a substantial population which will remain without a minimal level of service after current projects are completed. It is likely that the unserved population is located in more rural colonias or those far away from central water systems or wastewater treatment plants. In addition, as the colonias' population grow, the number of people in need of potable water and wastewater collection will surely rise. In order to provide service to the residents and colonias left unserved by current projects, more funding is needed.

Appendix A also shows that 30,860 colonia residents, or 8.7 percent, will still need water service, and 139,626 colonia residents, or 39.2 percent, will still need wastewater service when all planned EDAP projects are completed. The TWDB estimates the cost to serve these residents will be $424.6 million. Assuming that residents who need water are not the same as those who need wastewater, that these are two wholly different sets of people, this estimate computes to $2,491 per resident for infrastructure delivery. If we assume that all residents still needing water service also need wastewater service, the cost rises to $3,041 per capita. Both of these cost estimates are comparable with expenses reported thus far. If unserved colonias are mapped against served colonias, it is likely that those unserved are farther from central water and wastewater systems, which will further increase the cost of delivering service.

Another problem with cost estimation of those not included in current EDAP projects is that the TWDB needs database does not reflect colonias or individual households that are using alternative wastewater disposal technology. If a significant number of colonia households are using alternative technology, then it is possible that the number of households included in the cost-per-household calculation is actually too high. The projected costs to serve remaining needs will be much higher on a per household and per resident basis. Central wastewater collection and treatment is not the only adequate means of disposal. Septic tanks can be an appropriate and economical solution to wastewater disposal on a small scale, where lot size and terrain are compatible with design. When determining how many colonias, households and residents lack safe drinking water supply, this problem is not as relevant. Colonias have few means of accessing water other than through a water distributor, as high salinity in the groundwater of the Lower Rio Grande Basin prevents widespread use of wells for domestic drinking water.

The TWDB asserts that current and planned EDAP projects will deplete appropriated EDAP funds and that the State of Texas will need to allocate additional resources to serve those colonia residents left unaided by current and planned projects. In 1989, Texas voters approved $250 million in general obligation bonds to fund EDAP projects. Of that, $117.2 million has already been committed or spent on projects classified as "funded." The additional funding for these projects has come from other state and federal funding sources. A relatively insignificant amount of EDAP funding is loan funding, but repayment of these loans cannot be expected to provide enough money to meet the needs of those who will still be unserved after all appropriated funds are spent. The remaining

$133 million bonds will undoubtedly be spent serving projects in the "planning" stage. However, this does not resolve the problem of meeting the needs of the remaining 139,626 colonia residents. Leaving these residents without proper wastewater disposal systems may incur higher costs than providing the systems. Poor health and irregular school attendance could result in higher medical and welfare costs in the future. To avoid this, the State of Texas must decide whether to allocate new funding or to risk further negative health and environmental impacts and their associated future costs.

Recommendations

On the basis of this analysis, this report recommends that when the TWDB next performs an update to the 1992 Needs Survey, its analysts should compute separately all colonias which will have full access to water and wastewater facilities after the completion of all EDAP projects. However, a new update still will not be able to accurately predict which water and wastewater facilities will serve the colonias left without adequate facilities. Since water and wastewater facilities often operate as natural monopolies, regulated by state government and lacking competition, this report recommends that the Texas Legislature consider the establishment of statewide regulations that mandate utility provision of a minimum level of water and wastewater infrastructure as a prerequisite for sale of subdivided, platted land.

The legislature should consider amending the EDAP legislation to allow colonias to apply independently for EDAP assistance. The legislature should be aware that almost 44 percent of all colonias will not be served by current and planned EDAP projects. Some of these colonias may already have access to both water and wastewater facilities, but many are sure to be lacking. The process of obtaining a sponsor for EDAP projects is under no central control and does not establish a priority list of needs. Colonias should not be dependent on political subdivisions to obtain access to water and sewer lines. State, regional, and local governments and community-based organizations should increase outreach efforts to encourage closely-located groups of colonias to apply for funds, maximizing the potential for a single construction effort and minimizing operation and maintenance costs for individual colonias.

The Texas Legislature should consider the proliferation of colonias outside the Texas-Mexico border area. The TWDB has identified and planned projects in subdivisions characterized as colonias in East Texas. The legislature should take initiative to prevent further development of subdivisions lacking basic utility service. A statewide subdivision code should be implemented to promote additional low-income housing developments and to prevent the sale of lots and homes not equipped with access to central water and appropriate wastewater facilities.

Regardless of the accuracy of TWDB needs estimates, the Texas Legislature should consider appropriating additional EDAP funding for colonias with unmet needs. Population is growing, and Texas is becoming more urbanized. The provision of basic infrastructure will reduce the certain need of future financial assistance for medical services and will benefit more than the current population. As these areas grow, the cost to deliver water and wastewater service will grow. The state is encouraged to make the commitment now, before the cost becomes more prohibitive.

Notes

[1]Texas Water Development Board (TWDB), *Water and Wastewater Needs of Texas Colonias: 1995 Update* (Austin, Texas, February 1995), p. 8.

[2]Ibid., p. 7.

[3]C. Richard Bath, Janet M. Transki, and Roberto E. Villarreal, "The Politics of Water Allocation in El Paso County Colonias," *Journal of Borderlands Studies*, vol. 9, no. 1 (Spring 1994), p. 17.

[4]Lyndon B. Johnson School of Public Affairs, *Colonia Housing and Infrastructure: Current Population and Housing Characteristics, Future Growth, Housing, Water and Wastewater Needs*, Policy Research Project Preliminary Report (Austin, Texas, January 1996).

[5]Robert K. Holz and C. Shane Davies, *Third World Texas: Colonias in the Lower Rio Grand Valley*, Working Paper Series, no. 72 (Austin, Texas: Lyndon B. Johnson School of Public Affairs, August 1989), p. 13.

[6]Ibid.

[7]Ibid., p. 12.

[8]International City/County Management Association, *Texas and New Mexico Colonias: Barriers and Incentives for Local Government Involvement* (Washington, D.C., 1995), pp. 5-6.

[9]Ann Walthers, "Texas Colonias: On the Border of Misery" (Professional Report, Institute for Latin American Studies, The University of Texas at Austin, December 1988), p. 116.

[10]Texas Legislature, "Draft Report from El Paso Members of the Texas Legislature Regarding H.B. 1001," August 18, 1995, p. 6.

[11]Mark Killgore and David Eaton, *NAFTA Handbook for Water Resource Engineers* (Austin, Texas: U.S.-Mexican Policy Studies Program; and New York: American Society of Civil Engineers, 1995), pp. 28, 32.

[12]TWDB, *Water for Texas: Water and Wastewater Needs of Colonias in Texas* (Austin, Texas, October 1992), p. 1.

[13]Steve Mendoza, Engineering Specialist, Economically Distressed Areas Program, TWDB, Austin,

Texas, interview by Gina Briley, March 1996.

[14]Ibid.

[15]TWDB, "Colonia Plumbing Loan Program," Austin, Texas, February 1995 (pamphlet).

[16]TWDB, unpublished data provided to the PRP class, March 1996.

[17]Texas Department of Housing and Community Affairs (TDHCA), *State Low-Income Housing Plan and Annual Report* (Austin, Texas, 1995).

[18]Memorandum from Ruth Cedillo, Director, Community Development, TDHCA, to Executive Director Larry Paul Manley, TDHCA, Nov. 6, 1995.

[19]Ibid.

[20]TDHCA, *State Low-Income Housing Plan*, p. 274.

[21]Jorge Arroyo, Unit Chief, Colonias Planning Unit, Local and Regional Assistance Division, TWDB, Austin, Texas, interview by Sheila Cavanagh, February 5, 1997.

[22]Paco Valentin, Rural Economic and Community Development Agency, U.S. Department of Agriculture, Austin, Texas, interview by Sheila Cavanagh, March 21, 1996.

[23]Ibid.

[24]Roy Ramirez, Economic Development Representative, Economic Development Administration, U.S. Department of Commerce, Austin, Texas, interview by Sheila Cavanagh, March 21, 1996.

[25]Ibid.

[26]Ibid.

[27]Border Environment Cooperation Commission, *Draft Project List for Public Information*, Ciudad Juarez, Mexico (December 1996).

[28]Ibid.

[29]Edgar Tovilla, Border Environment Cooperation Commission, facsimile transmission of unpublished

data to Sheila Cavanagh, November 14, 1996.

[30]Sheila Cavanagh, "Right Purpose, Wrong Tools: NAFTA's Environmental Institutions and Texas Border Infrastructure" (Professional Report, Lyndon B. Johnson School of Public Affairs, May 1996), p. 95.

[31]TWDB, *Water for Texas*, 1992, p. 1.

[32]Curtis Johnson, Chief, Facility Needs Section, Planning Division, TWDB, Austin, Texas, interview by Gina Briley, February and March 1996.

[33]TWDB, *Water for Texas,* 1992, p. 3.

[34]Johnson interview, 1996

[35]Ibid.

[36]TWDB, *Water and Wastewater*, 1995, p. 2.

[37]Ibid., p. 2 and p. 5.

[38]TWDB, *Water for Texas*, 1992, p. 3.

[39]Ibid.

[40]Ibid.

[41]Ibid.

[42]Ibid., p.1.

[43]Johnson interview, 1996.

[44]Ibid.

[45]Mendoza interview, 1996.

[46]Ibid.

[47]Ibid.

[48]Ibid.

[49]Johnson interview, 1996.

[50]Ibid.

[51]Mendoza interview, 1996; and Johnson interview, 1996.

[52]Exiquio Salinas, *The Colonias Factbook: A Survey of Living Conditions in Rural Areas of South and West Texas Border Counties* (Austin, Texas: Texas Department of Human Services, 1988).

[53]Heidi Tran, Engineering Assistant, Economically Distressed Areas Program, TWDB, Austin, Texas, telephone interview by Gina Briley, April 1996.

Appendices

Appendix A. Texas Water Development Board Economically Distressed Areas Program

Table Notes and Sources

Column Title	Description	Source
Project Status	EDAP stage of infrastructure improvement project. Stages are: completed, under construction, plans and specifications, facilities planning, or colonias not served in EDAP.	Texas Water Development Board's "Economically Distressed Areas Program, Summary of EDAP Projects" as of February 15, 1996. Names of colonias in each project supplied by the Texas Water Development Board through unpublished data provided to the LBJ School of Public Affairs, University of Texas at Austin, April 2, 1996. Colonias not served by EDAP determined by process of elimination.
Project Name/Sponsor	Name of the applicant submitting plans for infrastructure improvement funds from the Economically Distressed Areas Program (EDAP)	Texas Water Development Board's "Economically Distressed Areas Program, Summary of EDAP Projects" as of February 15, 1996.
Name	Colonia name as designated by the Texas Water Development Board	Texas Water Development Board's Needs Database as provided to the LBJ School of Public Affairs, University of Texas at Austin, December, 1995.
Number of Residents	Estimated current colonia population.	Texas Water Development Board's Needs Database as provided to the LBJ School of Public Affairs, University of Texas at Austin, December, 1995.
Density	The average density of a household in the colonia.	Calculation : "Number of Residents" divided by "Number of Dwellings."
Number Served by Public Water	Colonia population that is currently served by a community water system (as defined by TNRCC standards).	Texas Water Development Board's Needs Database as provided to the LBJ School of Public Affairs, University of Texas at Austin, December, 1995.
Number Not Served by Public Water	Colonia population that is not currently served by a community water system.	Texas Water Development Board's Needs Database as provided to the LBJ School of Public Affairs, University of Texas at Austin, December, 1995.
Number Served by Central Wastewater	Colonia population that is currently served by a centralized wastewater treatment and collection system.	Texas Water Development Board's Needs Database as provided to the LBJ School of Public Affairs, University of Texas at Austin, December, 1995.

Table Notes and Sources continued

Column Title	Description	Source
Number Not Served by Central Wastewater	Colonia population that is currently not served by a centralized wastewater treatment and collection system.	Texas Water Development Board's Needs Database as provided to the LBJ School of Public Affairs, University of Texas at Austin, December, 1995.
Number Served by Project	The number of people in a colonia that can benefit from an EDAP project.	Either the total colonia population OR the sum of the number of people not served by public water and the number of people not served by central wastewater, *whichever is less.* Note: populations served by more than one EDAP project are only included in the number of people served in that one and only that one project. In addition, if a project is only designated for water service or wastewater service (as described in the TWDB's "Economically Distressed Areas Program, Summary of EDAP Projects" as of February 15, 1996), only the population not receiving that service is counted.
Number of Dwellings	The number of dwellings in a colonia.	Texas Water Development Board's Needs Database as provided to the LBJ School of Public Affairs, University of Texas at Austin, December, 1995.
Number of Occupied Lots	The number of lots or sites within the colonia that are occupied by homes or businesses.	Texas Water Development Board's Needs Database as provided to the LBJ School of Public Affairs, University of Texas at Austin, December, 1995.
Total Number of Lots	The total number of recorded lots in the colonia. If unknown, this number "defaults" to the number of occupied lots or sites.	Texas Water Development Board's Needs Database as provided to the LBJ School of Public Affairs, University of Texas at Austin, December, 1995.
Project Status	EDAP stage of infrastructure improvement project. Stages are: completed, under construction, plans and specifications, or facilities planning.	Texas Water Development Board's "Economically Distressed Areas Program, Summary of EDAP Projects" as of February 15, 1996. Names of colonias in each project supplied by the Texas Water Development Board through unpublished data provided to the LBJ School of Public Affairs, University of Texas at Austin, April 2, 1996. Colonias not served by EDAP determined by process of elimination.

Table Notes and Sources continued

Column Title	Description	Source
Project Name / Sponsor	Name of the applicant submitting plans for infrastructure improvement funds from the Economically Distressed Areas Program (EDAP).	Texas Water Development Board's "Economically Distressed Areas Program, Summary of EDAP Projects" as of February 15, 1996. Names of colonias in each project supplied by the Texas Water Development Board through unpublished data provided to the LBJ School of Public Affairs, University of Texas at Austin, April 2, 1996.
Estimated Total Project Costs	The estimated total costs of infrastructure improvement projects, including the purchase of water rights and planning costs if in the facilities planning stage.	Texas Water Development Board's "Economically Distressed Areas Program, Summary of EDAP Projects" as of February 15, 1996.
Project Cost Per Capita	The project cost per capita based on all residents of a colonia, not including those that have been served by a previously listed project(s).	Calculation: "Estimated Total Project Costs" divided by "Number of Residents." Note: For colonias which have been served by a previously listed project, colonia population is not included in the total sum for the project's cost calculations.
Project Cost Per Capita Served	The project cost per capita based on the total number of people served by the project.	Calculation: "Estimated Total Project Costs" divided by "Number Served by Project."
Project Cost Per Dwelling	The project cost per dwelling based on the total number of dwellings in each colonia, not including those dwellings that have been served by a previously listed project(s).	Calculation: "Estimated Total Project Costs" divided by "Number of Dwellings." Note: For colonias which have been served by a previously listed project, number of dwellings in that colonia is not included in the total sum for the project's cost calculations.
Project Cost Per Dwelling (Density)	The project cost per dwelling based on the estimated number of homes to be served, as calculated using average density for the colonia applied to the number of people served by the project, not including those dwellings that have been served by a previously listed project.	Calculation: "Estimated total project costs" divided by the quantity ["Number Served by Project" divided by "Density"].
Project Cost per Lot at Buildout	The project cost per lot in a colonia or group of colonias.	Calculation: "Estimated Total Project Costs" divided by "Total Number of Lots."

Table A.1
Bee County Project Summary

Project Status	Project Name / Sponsor	Colonia Name	# of Residents	Density	# Svd. by Public Water	# Not Svd. by Public Water	# Svd. by Central Waste-water	# Not Svd. by Central Waste-water	# Svd. by Project	# of Dwellings	# of Occupied Lots	Total # of Lots
Colonias Not Served in EDAP												
		Blueberry Hill	375	3.000	375	0	0	375		125	125	125
		Skidmore	300	3.000	0	300	0	300		100	100	100
		Tuleta	120	3.000	0	120	0	120		40	40	40
		Tynan	474	3.000	474	0	0	474		158	158	158
NOT SERVED BY EDAP - TOTAL BEE COUNTY		**Number of Colonias: 4**	**1,269**	**3.000**	**849**	**420**	**0**	**1,269**	**0**	**423**	**423**	**423**

Table A.2
Brooks County Project Summary

Project Status	Project Name / Sponsor	Colonia Name	# of Residents	Density	# Svd. by Public Water	# Not Svd. by Public Water	# Svd. by Central Waste-water	# Not Svd. by Central Waste-water	# Svd. by Project	# of Dwellings	# of Occupied Lots	Total # of Lots
Colonias Not Served in EDAP												
		Airport Road Subdivision	90	3.000	90	0	0	90		30	30	30
		Belmares	32	3.200	0	32	0	32		10	10	10
		Cantu Addition	55	2.500	55	0	55	0		22	22	22
		Encino	200	2.857	0	200	0	200		70	70	70
		La Parrita	135	3.000	100	35	0	135		45	45	45
		Rush Addition	35	3.182	0	35	0	35		11	11	11
		Whisler Addition	65	2.600	65	0	65	0		25	25	25
NOT SERVED BY EDAP - TOTAL BROOKS COUNTY		**Number of Colonias:** 7	**612**	**2.873**	**310**	**302**	**120**	**492**	**0**	**213**	**213**	**213**

Table A.3
Cameron County Project Summary

Project Status	Project Name / Sponsor	Colonia Name	# of Residents	Density	# Svd. by Public Water	# Not Svd. by Public Water	# Svd. by Central Waste-water	# Not Svd. by Central Waste-water	# Svd. by Project	# of Dwellings	# of Occupied Lots	Total # of Lots
Completed												
	City of Brownsville (Cameron Park)											
		Cameron Park	4,398	5.841	4,398	0	0	4,398	4,398	753	753	1,624
	TOTAL - City of Brownsville (Cameron Park)		**4,398**	**5.841**	**4,398**	**0**	**0**	**4,398**	**4,398**	**753**	**753**	**1,624**
	City of Brownsville (Hacienda Gardens)											
		Hacienda Gardens	308	4.219	308	0	0	308	308	73	73	117
	TOTAL - City of Brownsville (Hacienda Gardens)		**308**	**4.219**	**308**	**0**	**0**	**308**	**308**	**73**	**73**	**117**
Plans and Specifications												
	Olimito WSC											
		Olmito	3,790	3.500	3,790	0	0	3,790	3,790	1,083	1,083	1,083
	TOTAL - Olimito WSC		**3,790**	**3.500**	**3,790**	**0**	**0**	**3,790**	**3,790**	**1,083**	**1,083**	**1,083**
Facilities Planning												
	Cameron County (Valle Hermosa & Valle Escondido)											
		Valle Escondido	137	4.893	137	0	0	137		28	28	56
		Valle Hermosa	64	4.923	64	0	0	64		13	13	26
	TOTAL - Cameron County (Valle Hermosa & Valle Escondido)		201	4.902	201	0	0	201	201	41	41	82
	Rio Hondo											
		Bullis Subdivision	91	3.792	91	0	0	91		24	24	24
		Jones Addition	91	3.792	91	0	0	91		24	24	24
		Lopez Subdivision	60	3.750	60	0	0	60		16	16	16
		Tatum Addition	11	3.667	11	0	0	11		3	3	3
		West Addition	83	3.773	83	0	0	83		22	22	22
	TOTAL - Rio Hondo		**336**	**3.775**	**336**	**0**	**0**	**336**	**336**	**89**	**89**	**89**

Continued

Table A.3 (continued)
Cameron County Project Summary

Project Status	Project Name / Sponsor	Colonia Name	# of Residents	Density	# Svd. by Public Water	# Not Svd. by Public Water	# Svd. by Central Waste-water	# Not Svd. by Central Waste-water	# Svd. by Project	# of Dwellings	# of Occupied Lots	Total # of Lots
Facilities Planning (continued)	Rural Planning/Cameron County											
		Arroyo Gardens	97	3.593	97	0	0	97		27	17	26
		Bishop Subdivision	123	4.731	123	0	0	123		26	22	26
		Del Mar Heights	344	3.475	344	0	0	344		99	121	1,704
		Glenwood Acres Subd.	172	4.914	172	0	0	172		35	35	45
		La Tina Ranch	519	4.896	519	0	0	519		106	106	135
		Laguna Escondido	88	4.190	88	0	0	88		21	21	99
		Las Yescas	221	4.911	221	0	0	221		45	45	62
		Lozano	534	4.899	534	0	0	534		109	109	139
		Orason/Chula Vista/Shoemkr	235	4.896	235	0	0	235		48	48	95
		Palmer	162	3.600	162	0	0	162		45	43	58
		Rangerville	105	3.889	105	0	0	105		27	22	28
		Rangerville Estates	280	4.516	280	0	0	280		62	62	62
		San Vicente	81	3.682	81	0	0	81		22	16	20
		Santa Elena	150	3.846	150	0	0	150		39	37	51
	TOTAL -Rural Planning/Cameron County		**3,111**	**4.376**	**3,111**	**0**	**0**	**3,111**	**3,111**	**711**	**704**	**2,550**
	San Benito											
		Camino Angosto	236	4.453	236	0	0	236		53	53	53
		East Expressway 83/77	60	4.286	60	0	0	60		14	14	14
		East Stenger Street	116	4.462	0	116	0	116		26	26	26
		Leal Subdivision	185	4.405	185	0	0	185		42	42	44
		South Ratliff Street	18	2.250	0	18	0	18		8	8	8
		Yost Road	116	4.462	116	0	0	116		26	26	26
	TOTAL - San Benito		**731**	**4.325**	**597**	**134**	**0**	**731**	**731**	**169**	**169**	**171**
	Urban Regional Wastewater Planning/Cameron County											
		Arroyo Colorado Estates	791	4.853	791	0	0	791		163	150	460

Continued

Table A.3 (continued)
Cameron County Project Summary

Project Status	Project Name / Sponsor	Colonia Name	# of Residents	Density	# Svd. by Public Water	# Not Svd. by Public Water	# Svd. by Central Waste-water	# Not Svd. by Central Waste-water	# Svd. by Project	# of Dwellings	# of Occupied Lots	Total # of Lots
Facilities Planning (continued)		Barrio Subdivision	136	4.857	136	0	0	136		28	28	28
		Bautista	565	4.788	565	0	0	565		118	118	192
		Bettie Acres & Harris Tract	564	5.036	564	0	0	564		112	112	377
		Casa del Rey	415	4.560	415	0	0	415		91	91	169
		Combes	2,366	4.498	2,366	0	0	2,366		526	526	526
		Eggers	138	4.600	138	0	0	138		30	30	66
		Illinois Heights	186	4.133	186	0	0	186		45	45	131
		Indian Lake	400	4.444	400	0	0	400		90	90	90
		La Coma	72	3.000	72	0	0	72		24	24	31
		Laguna Escondido	Already served under Rural Planning/Cameron County Project.									
		Lasana	168	3.907	168	0	0	168		43	43	286
		Laureles	1,453	4.860	1,453	0	0	1,453		299	226	226
		Los Cuates	191	4.897	191	0	0	191		39	39	77
		Lourdes Street	175	5.000	175	0	0	175		35	35	61
		Palacios	58	6.444	58	0	0	58		9	9	66
		Paredes Estates	170	4.857	170	0	0	170		35	35	66
		Primera	2,352	4.497	2,352	0	0	2,352		523	523	523
		Saldivar II (Unknown Hockaday)	192	6.000	192	0	0	192		32	32	50
		Saldivar-Central Estates	396	5.211	396	0	0	396		76	76	87
		Unknown Boca Chica & Medford	155	5.536	155	0	0	155		28	28	33
		Unknown Houston Rd East	53	4.818	53	0	0	53		11	11	55
		Unknown Travis & Vermillion Rds	101	4.810	101	0	0	101		21	21	55
		Unknown (Travis Rd East)	108	4.909	0	108	0	108		22	22	28

Continued

Table A.3 (continued)
Cameron County Project Summary

Project Status	Project Name / Sponsor	Colonia Name	# of Residents	Density	# Svd. by Public Water	# Not Svd. by Public Water	# Svd. by Central Waste-water	# Not Svd. by Central Waste-water	# Svd. by Project	# of Dwellings	# of Occupied Lots	Total # of Lots
Facilities Planning (continued)		Unnamed (Hwy 802)	182	4.136	182	0	0	182		44	44	110
	TOTAL - Urban Regional Wastewater Planning/Cameron County		**11,387**	**4.659**	**11,279**	**108**	**0**	**11,387**	**11,387**	**2,444**	**2,358**	**3,793**
EDAP TOTAL CAMERON COUNTY		**Number of Colonias: 55**	**24,262**	**4.524**	**24,020**	**242**	**0**	**24,262**	**24,262**	**5,363**	**5,270**	**9,509**
Colonias Not Served in EDAP												
		21 Subdivision	44	4.889	44	0	0	44		9	9	18
		511 Crossroads	123	4.920	123	0	0	123		25	25	50
		Alabama/Arkansas (La Coma)	515	4.905	515	0	0	515		105	105	209
		Alto Real Subdivision	115	6.389	115	0	0	115		18	16	17
		Arroyo Alto	183	4.463	183	0	183	0		41	41	41
		Bixby Townsite	51	6.375	51	0	0	51		8	8	14
		Bluetown	456	4.903	456	0	0	456		93	93	118
		Carricitos-Landrum	216	4.909	216	0	0	216		44	44	56
		Coronado	152	4.903	152	0	0	152		31	31	62
		Dakota Mobile Homes	294	4.900	294	0	294	0		60	60	119
		East Cantu Country Est	186	7.440	0	186	0	186		25	24	29
		East Cantu Road Subd	64	6.400	0	64	0	64		10	10	20
		El Calaboz	181	4.892	181	0	0	181		37	37	73
		El Venadito	225	4.891	225	0	0	225		46	46	59
		Encantada/El Ranchito	1,289	4.901	1,289	0	0	1,289		263	263	335
		Fred Adams & Juarez	542	4.517	542	0	542	0		120	120	120
		Iglesia Antiqua	162	4.909	162	0	0	162		33	33	42

Continued

Table A.3 (continued)
Cameron County Project Summary

Project Status	Project Name / Sponsor	Colonia Name	# of Residents	Density	# Svd. by Public Water	# Not Svd. by Public Water	# Svd. by Central Waste-water	# Not Svd. by Central Waste-water	# Svd. by Project	# of Dwelling s	# of Occupied Lots	Total # of Lots
Colonias Not Served in EDAP (continued)		Jaime Lake	98	4.900	98	0	98	0		20	20	40
		La Paloma	676	4.899	676	0	0	676		138	138	176
		Lago	392	4.900	392	0	0	392		80	80	142
		Las Palmas	622	4.898	622	0	0	622		127	127	225
		Las Rusias	258	4.526	258	0	0	258		57	57	57
		Los Cuates (South)	96	3.556	96	0	0	96		27	27	77
		Los Indios	549	4.902	549	0	0	549		112	112	143
		Los Ranchitos	318	4.609	318	0	0	318		69	69	69
		Rancho Solis	192	6.400	192	0	0	192		30	27	42
		Ratamosa Subd - Block 1	154	6.417	0	154	0	154		24	24	41
		Rice Tracts	132	4.889	132	0	0	132		27	27	48
		San Pedro	730	4.899	730	0	0	730		149	149	296
		Santa Maria	1,161	4.899	1,161	0	0	1,161		237	237	471
		Santa Rosa # 1	413	4.976	0	413	0	413		83	83	83
		Santa Rosa # 2	220	4.231	0	220	0	220		52	52	52
		Santa Rosa # 3	170	3.469	0	170	0	170		49	49	49
		Santa Rosa # 4	133	4.030	0	133	0	133		33	33	33
		Santa Rosa # 5	111	3.581	0	111	0	111		31	31	31
		Santa Rosa # 6	102	3.517	0	102	0	102		29	29	29
		Santa Rosa # 7	127	4.536	0	127	0	127		28	28	28
		Santa Rosa # 8	105	4.773	0	105	0	105		22	22	22
		Santa Rosa # 9	47	2.474	0	47	0	47		19	19	19
		Santa Rosa # 10	64	4.571	0	64	0	64		14	14	14
		Santa Rosa # 11	41	3.727	0	41	0	41		11	11	11
		Santa Rosa # 12	44	3.667	0	44	0	44		12	12	12
		Santa Rosa # 13	22	2.444	0	22	0	22		9	9	9
		Santa Rosa # 14	40	6.667	0	40	0	40		6	6	6
		Santa Rosa # 15	32	6.400	0	32	0	32		5	5	5

Continued

Table A.3 (continued)
Cameron County Project Summary

Project Status	Project Name / Sponsor	Colonia Name	# of Residents	Density	# Svd. by Public Water	# Not Svd. by Public Water	# Svd. by Central Waste-water	# Not Svd. by Central Waste-water	# Svd. by Project	# of Dwellings	# of Occupied Lots	Total # of Lots
Colonias Not Served in EDAP (continued)		Santa Rosa # 16	20	4.000	0	20	0	20		5	5	5
		Solis Road	83	7.545	83	0	0	83		11	11	12
		Stardust	117	4.034	0	117	0	117		29	29	29
		Stuart Subdivision	990	4.901	990	0	990	0		202	202	400
		Unknown (Oklahoma Ave)	115	3.710	115	0	0	115		31	31	50
		Unknown (Rabb Road)	142	4.897	142	0	0	142		29	29	58
		Unnamed C	132	4.889	132	0	0	132		27	27	54
		Villa Cavazos	201	4.902	201	0	0	201		41	41	81
		Villa Nueva	402	4.902	402	0	0	402		82	82	163
		Villa Pancho	304	4.903	304	0	0	304		62	62	123
		Wymore Subd - Aurora Ave	122	6.421	122	0	0	122		19	19	29
		Yznaga Subdivision	102	6.375	0	102	0	102		16	16	21
NOT SERVED BY EDAP - TOTAL CAMERON COUNTY		**Number of Colonais: 57**	**14,577**	**4.824**	**12,263**	**2,314**	**2,107**	**12,470**	**53,230**	**3,022**	**3,016**	**4,637**

Table A.4
Coryell County Project Summary

Project Status	Project Name / Sponsor	Colonia Name	# of Residents	Density	# Svd. by Public Water	# Not Svd. by Public Water	# Svd. by Central Waste-water	# Not Svd. by Central Waste-water	# Svd. by Project	# of Dwellings	# of Occupied Lots	Total # of Lots
Facilities Planning												
	Copperas Cove											
		Big Valley Ranchettes	174	3.283	174	0	0	174		53	53	119
		Bradford Oaks	81	3.240	81	0	0	81		25	25	30
		Cove Acres	33	3.300	33	0	0	33		10	10	10
		Gibson Street	467	2.638	467	0	0	467		177	177	86
		Oak Hill	66	2.640	66	0	0	66		25	25	23
		Pecan Cove	165	3.300	165	0	0	165		50	50	76
		Ritter Street	42	3.000	42	0	0	42		14	14	28
		Tanglewood Estates	42	3.231	42	0	0	42		13	13	35
		Twin Hills Ranchetttes	75	3.261	75	0	0	75		23	23	21
		Willow Springs # 1	137	3.605	137	0	0	137		38	38	72
		Willow Springs # 2	177	3.612	177	0	0	177		49	49	75
		Woodland Pk/Lutheran Church	153	3.255	153	0	0	153		47	47	47
	TOTAL - Copperas Cove		**1,612**	**3.076**	**1,612**	**0**	**0**	**1,612**	**1,612**	**524**	**524**	**622**
	Gatesville											
		Fort Gates (Unserved Areas)	662	2.691	662	0	0	662		246	246	246
		Gatesville (Unserved Areas)	207	2.688	207	0	0	207		77	77	77
	TOTAL - Gatesville		**869**	**2.690**	**869**	**0**	**0**	**869**	**869**	**323**	**323**	**323**
EDAP TOTAL CORYELL COUNTY		**Number of Colonias: 14**	**2,481**	**2.929**	**2,481**	**0**	**0**	**2,481**	**2,481**	**847**	**847**	**945**

Table A.5
Dimmit County Project Summary

Project Status	Project Name / Sponsor	Colonia Name	# of Residents	Density	# Svd. by Public Water	# Not Svd. by Public Water	# Svd. by Central Waste-water	# Not Svd. by Central Waste-water	# Svd. by Project	# of Dwellings	# of Occupied Lots	Total # of Lots
Facilities Planning												
	Dimmit County											
		Asherton	1,610	3.096	1,610	0	1,610	0		520	520	1,900
		Carrizo Hills	514	3.295	0	514	0	514		156	156	200
		Catarina	238	3.306	160	78	0	238		72	72	2,500
	TOTAL - Dimmit County		**2,362**	**3.158**	**1,770**	**592**	**1,610**	**752**	**1,344***	**748**	**748**	**4,600**
		* Based on the assumption that people who need water are not the same people who need wastewater.										
EDAP TOTAL DIMMIT COUNTY		**Number of Colonias: 3**	**2,362**	**3.158**	**1,770**	**592**	**1,610**	**752**	**1,344****	**748**	**748**	**4,600**
		** Based on the assumption that people who need water are not the same people who need wastewater.										
Colonias Not Served in EDAP												
		Big Wells	1,557	4.461	1,557	0	1,357	200		349	349	2,000
		Brundage	32	4.000	0	32	0	32		8	8	1,500
		Espantosa	188	4.000	0	188	0	188		47	47	150
NOT SERVED BY EDAP - TOTAL DIMMIT COUNTY		**Number of Colonias: 3**	**1,777**	**4.399**	**1,557**	**220**	**1,357**	**420**	**0**	**404**	**404**	**3,650**

Table A.6
Duval County Project Summary

Project Status	Project Name / Sponsor	Colonia Name	# of Residents	Density	# Svd. by Public Water	# Not Svd. by Public Water	# Svd. by Central Waste-water	# Not Svd. by Central Waste-water	# Svd. by Project	# of Dwellings	# of Occupied Lots	Total # of Lots
Colonias Not Served in EDAP												
		Cadena	100	2.857	100	0	0	100		35	35	70
NOT SERVED BY EDAP - TOTAL DUVAL COUNTY		**Number of Colonias: 1**	**100**	**2.857**	**100**	**0**	**0**	**100**	**0**	**35**	**35**	**70**

Table A.7
Edwards County Project Summary

Project Status	Project Name / Sponsor	Colonia Name	# of Residents	Density	# Svd. by Public Water	# Not Svd. by Public Water	# Svd. by Central Waste-water	# Not Svd. by Central Waste-water	# Svd. by Project	# of Dwellings	# of Occupied Lots	Total # of Lots
Colonias Not Served in EDAP												
		Rocksprings	1,321	2.642	1,321	0	1,321	0		500	500	500
NOT SERVED BY EDAP - TOTAL EDWARDS COUNTY		**Number of Colonias: 1**	**1,321**	**2.642**	**1,321**	**0**	**1,321**	**0**	**0**	**500**	**500**	**500**

Table A.8
El Paso County Project Summary

Project Status	Project Name / Sponsor	Colonia Name	# of Residents	Density	# Svd. by Public Water	# Not Svd. by Public Water	# Svd. by Central Waste-water	# Not Svd. by Central Waste-water	# Svd. by Project	# of Dwellings	# of Occupied Lots	Total # of Lots
Completed												
	EPCLVWDA (Socorro-Bauman Water Project)											
		Aldama Est.	266	5.429	0	266	0	266		49	49	67
		Bauman	1,127	5.418	0	1,127	0	1,127		208	208	256
		Belen Plaza	271	5.420	0	271	0	271		50	50	70
		Rio Rancho	168	5.419	0	168	0	168		31	31	47
		San Agustin	228	5.429	0	228	0	228		42	42	54
		San Ysidro	412	5.421	0	412	0	412		76	76	95
	TOTAL - EPCLVWDA (Socorro-Bauman Water Project)		**2,472**	**5.421**	**0**	**2,472**	**0**	**2,472**	**2,472**	**456**	**456**	**589**
	EPCWCID											
		Westway	3,214	4.720	3,214	0	1,120	2,094	2,094	681	681	1,305
	TOTAL - EPCWCID		**3,214**	**4.720**	**3,214**	**0**	**1,120**	**2,094**	**2,094**	**681**	**681**	**1,305**
Plans and Specifications												
	City of El Paso											
		Westway	Already Served under EPCWCID Project.									
	TOTAL - City of El Paso		**0**	**0.000**	**0**	**0**	**0**	**0**	**0**	**0**	**0**	**0**
	EPLVWDA Socorro Phase II											
		Algodon										
		Bagge Est.	531	5.418	531	0	0	531		98	98	124
		Calcutta	38	5.429	38	0	0	38		7	7	8
		Carreta										
		Delip Subd.	1,518	5.421	1,518	0	0	1,518		280	280	339
		Ellen Park	385	5.423	385	0	0	385		71	71	83
		Frank (partial)	65	5.417	0	65	0	65		12	12	17
		Grijalva Gardens	813	5.420	813	0	0	813		150	150	171
		Gurdev Subd.	976	5.422	976	0	0	976		180	180	224
		Hillcrest Manor	54	4.500	54	0	0	54		12	12	267

Continued

Table A.8 (continued)
El Paso County Project Summary

Project Status	Project Name / Sponsor	Colonia Name	# of Residents	Density	# Svd. by Public Water	# Not Svd. by Public Water	# Svd. by Central Waste-water	# Not Svd. by Central Waste-water	# Svd. by Project	# of Dwellings	# of Occupied Lots	Total # of Lots
Plans and Specifications (continued)		La Fuente	157	5.414	157	0	0	157		29	29	39
		La Junta	293	5.426	293	0	0	293		54	54	61
		Monterosales Sub.	455	5.417	455	0	0	455		84	84	93
		Moon Addn.	1,306	5.419	1,306	0	0	1,306		241	241	275
		North Loop Acres	244	5.422	244	0	0	244		45	45	51
		Rio Rancho	Already served under EPCLVWDA (Socorro-Bauman Water Project).									
		Rio Vista	428	5.418	428	0	0	428		79	79	84
		San Augustin	Already served under EPCLVWDA (Socorro-Bauman Water Project).									
		Socorro Village	146	5.407	146	0	0	146		27	27	37
		Spanish Trail	569	5.419	569	0	0	569		105	105	122
		Sunshine	87	5.438	87	0	0	87		16	16	17
	TOTAL - EPLVWDA Socorro Phase II		**8,065**	**5.413**	**8,000**	**65**	**0**	**8,065**	**8,065**	**1,490**	**1,490**	**2,012**
	EPCLVWDA Socorro/San Elizario - Phase III											
		Adobe										
		Alameda Est.	883	5.417	883	0	0	883		163	163	190
		Aldama Est.	Already served under EPCLVWDA (Socorro-Bauman Water Project).									
		Aljo Est.	634	5.419	0	634	0	634		117	117	129
		Angie Subd.	125	5.435	125	0	0	125		23	23	25
		Bauman	Already served under EPCLVWDA (Socorro-Bauman Water Project).									
		Belen Plaza	Already served under EPCLVWDA (Socorro-Bauman Water Project).									
		Brinkman Addition	168	5.419	0	168	0	168		31	31	41
		Bufford View										
		Burbridge Acres	455	5.417	0	455	0	455		84	84	111
		Campo Bello Estates										
		Colonia De Las Azaleas	206	5.421	0	206	0	206		38	38	87
		Colonia Del Rio	878	5.420	878	0	0	878		162	162	342
		Country Green	1,241	5.419	1,241.	0	0	1,241		229	229	273
		El Campestre (partial)	1,003	5.422	1,003	0	0	1,003		185	185	237
		Frank-Anita Est.	146	5.407	0	146	0	146		27	27	60
		Friedman	2,461	5.421	2,461	0	0	2,461		454	454	598

Continued

Table A.8 (continued)
El Paso County Project Summary

Project Status	Project Name / Sponsor	Colonia Name	# of Residents	Density	# Svd. by Public Water	# Not Svd. by Public Water	# Svd. by Central Waste-water	# Not Svd. by Central Waste-water	# Svd. by Project	# of Dwellings	# of Occupied Lots	Total # of Lots
Plans and Specifications (continued)		Gloria Elena	70	5.385	0	70	0	70		13	13	34
		Glorieta	179	5.424	179	0	0	179		33	33	42
		Gonzalez	87	5.438	0	87	0	87		16	16	36
		Horizon Ind Park										
		Jones	125	5.435	0	125	0	125		23	23	63
		La Jolla	585	5.417	0	585	0	585		108	108	123
		Las Aves	163	4.528	163	0	0	163		36	36	51
		Las Milpas	304	5.429	0	304	0	304		56	56	62
		Lewis Subd.	260	5.417	260	0	0	260		48	48	48
		Lordsville	114	5.429	49	65	0	114		21	21	29
		Lynn Park	900	5.422	900	0	0	900		166	166	190
		Madrilena	119	5.409	119	0	0	119		22	22	22
		Mary Lou Park	585	5.417	585	0	0	585		108	108	123
		Plaza Bernal	304	5.429	304	0	0	304		56	56	76
		Poole	466	5.419	0	466	0	466		86	86	135
		Quail Mesa	103	5.421	103	0	0	103		19	16	16
		Rancho Miravel	195	5.417	0	195	0	195		36	36	53
		Rio Posado Est.	168	5.419	0	168	0	168		31	31	49
		Roseville	640	5.424	0	640	0	640		118	118	161
		San Paulo	217	5.425	0	217	0	217		40	37	37
		San Ysidro	Already served under EPCLVWDA (Socorro-Bauman Water Project).									
		Santa Martina	385	5.423	0	385	0	385		71	71	76
		Sun Haven	27	5.400	0	27	0	27		5	5	79
		Sylvia Andrea	314	5.414	0	314	0	314		58	58	53
		Valle Real	168	5.419	0	168	0	168		31	31	51
		Valle Villa	439	5.420	439	0	0	439		81	81	108
		Villa Espana	287	5.415	287	0	0	287		53	53	60
		Villalobos	417	5.416	417	0	0	417		77	77	99
		Vinedo Acres	575	5.425	0	575	0	575		106	106	141
		Wilton Acres	141	5.423	141	0	0	141		26	26	26

Continued

Table A.8 (continued)
El Paso County Project Summary

Project Status	Project Name / Sponsor	Colonia Name	# of Residents	Density	# Svd. by Public Water	# Not Svd. by Public Water	# Svd. by Central Waste-water	# Not Svd. by Central Waste-water	# Svd. by Project	# of Dwellings	# of Occupied Lots	Total # of Lots
Plans and Specifications (continued)		Wiseman	244	5.422	0	244	0	244		45	45	55
	TOTAL - EPCLVWDA Socorro/San Elizario - Phase III		**16,781**	**5.410**	**10,537**	**6,244**	**0**	**16,781**	**16,781**	**3,102**	**3,096**	**4,191**
	Homestead MUD											
		Deerfield Park	1,355	4.502	1,355	0	0	1,355		301	301	454
		Desert Glen	302	4.507	302	0	0	302		67	67	97
		Homestead Homes	455	4.505	455	0	0	455		101	101	138
		Homestead Meadows So.	1,755	4.500	1,755	0	0	1,755		390	390	572
		Las Casitas	567	4.500	567	0	0	567		126	126	218
		Las Quintas	401	4.506	401	0	0	401		89	89	130
		Southwest Est.	194	4.512	194	0	0	194		43	43	65
	TOTAL - Homestead MUD		**5,029**	**4.502**	**5,029**	**0**	**0**	**5,029**	**5,029**	**1,117**	**1,117**	**1,674**
Facilities Planning												
	City of El Paso (Canutillo)											
		Canutillo	2,152	4.719	2,152	0	0	2,152		456	456	644
		Hermosa Vista	198	6.188	198	0	0	198		32	32	48
	TOTAL - City of El Paso (Canutillo)		**2,350**	**4.816**	**2,350**	**0**	**0**	**2,350**	**2,350**	**488**	**488**	**692**
	Eastside Montana											
		Desert Meadows Est.	221	4.510	221	0	0	221		49	49	252
		Desert Oasis	468	4.500	468	0	0	468		104	104	104
		East Wind Est.	297	4.500	297	0	0	297		66	66	68
		Fern Village	162	4.500	162	0	0	162		36	36	71
		Flamingo	90	4.500	90	0	0	90		20	20	126
		Hillcrest Est.	168	5.419	168	0	0	168		31	31	55
		Homestead Meadows	990	4.500	990	0	0	990		220	220	394
		John Michael Est.	77	4.529	77	0	0	77		17	17	30
		May Estates	59	4.538	59	0	0	59		13	13	26
		McCracken Estates	338	4.507	338	0	0	338		75	75	88
		Montana East	41	4.556	41	0	0	41		9	9	103
		Montana Land Est.	252	4.500	252	0	0	252		56	56	78

Continued

Table A.8 (continued)
El Paso County Project Summary

Project Status	Project Name / Sponsor	Colonia Name	# of Residents	Density	# Svd. by Public Water	# Not Svd. by Public Water	# Svd. by Central Waste-water	# Not Svd. by Central Waste-water	# Svd. by Project	# of Dwellings	# of Occupied Lots	Total # of Lots
Facilities Planning (continued)		Montana Vista Est.	261	4.500	261	0	0	261		58	58	61
		Paso View	428	4.505	428	0	0	428		95	95	160
		Paso View West	171	4.500	171	0	0	171		38	38	38
		Square Dance	1,051	5.418	1,051	0	0	1,051		194	194	494
		Turf Est.	765	4.500	765	0	0	765		170	170	170
		Vista De Lomas	158	4.514	158	0	0	158		35	35	129
		Vista Del Este	1,058	4.502	1,058	0	0	1,058		235	235	330
		Vista Montana	275	4.508	275	0	0	275		61	61	61
	TOTAL - Eastside Montana		**7,330**	**4.633**	**7,330**	**0**	**0**	**7,330**	**7,330**	**1,582**	**1,582**	**2,838**
	EPLVWDA (Las Azaleas)											
		Colonia De Las Azaleas	Already served by EPCLVWDA Socorro/San Elizario - Phase III.									
	TOTAL - EPLVWDA (Las Azaleas)		**0**	**0.000**	**0**	**0**	**0**	**0**	**0**	**0**	**0**	**0**
	Tornillo WSC											
		Tornillo WSC	1,133	5.421	1,133	0	0	1,133		209	209	241
	TOTAL - Tornillo WSC		**1,133**	**5.421**	**1,133**	**0**	**0**	**1,133**	**1,133**	**209**	**209**	**241**
EDAP TOTAL EL PASO COUNTY		**Number of Colonias: 100**	**46,374**	**5.082**	**37,593**	**8,781**	**1,120**	**45,254**	**45,254**	**9,125**	**9,119**	**13,542**
Colonias Not Served in EDAP												
		Adelante Est.	80	4.706	80	0	0	80		17	17	24
		Agua Dulce	1,530	4.500	1,530	0	0	1,530		340	340	790
		Alvarez	38	5.429	0	38	0	38		7	7	7
		Ascension Park	271	5.420	0	271	0	271		50	50	128
		Athena West	163	5.433	0	163	0	163		30	30	155
		Bejar Est.	54	5.400	0	54	0	54		10	10	40
		Bill Bart Road	27	4.500	0	27	0	27		6	6	6
		Bosque Bonito	472	5.425	472	0	0	472		87	87	87
		Bovee Road Tract	65	5.417	0	65	0	65		12	12	12
		Camino Barrial	54	5.400	0	54	0	54		10	10	10

Continued

Table A.8 (continued)
El Paso County Project Summary

Project Status	Project Name / Sponsor	Colonia Name	# of Residents	Density	# Svd. by Public Water	# Not Svd. by Public Water	# Svd. by Central Waste-water	# Not Svd. by Central Waste-water	# Svd. by Project	# of Dwellings	# of Occupied Lots	Total # of Lots
Colonias Not Served in EDAP (continued)		Cattlemens North Rancho	16	5.333	0	16	0	16		3	3	131
		Clint	2,011	5.420	2,011	0	0	2,011		371	371	600
		Cochran Estates	65	5.417	0	65	0	65		12	12	12
		College Park	163	5.433	0	163	0	163		30	30	30
		Colonia De Las Dalias	1,127	5.418	0	1,127	0	1,127		208	208	253
		Connington	130	5.417	0	130	0	130		24	24	37
		Cotton Valley Estates	271	5.420	0	271	0	271		50	50	50
		Cuna Del Valle	537	5.424	537	0	0	537		99	99	127
		Dairyland	249	5.413	0	249	0	249		46	46	80
		Dindinger Road	450	5.422	0	450	0	450		83	83	83
		E & L	108	4.500	108	0	0	108		24	20	20
		East Clint	98	5.444	0	98	0	98		18	18	25
		El Gran Ville	1,051	5.418	0	1,051	0	1,051		194	194	216
		Friedman Est. 2	2,493	5.420	0	2,493	0	2,493		460	460	542
		Green Acres	76	4.750	76	0	0	76		16	16	48
		Hacienda Del Valle	60	5.455	60	0	0	60		11	11	26
		Hacienda Real	1,089	5.418	0	1,089	0	1,089		201	201	229
		La Union Est.	61	4.692	61	0	0	61		13	13	18
		Las Colonias	656	5.421	656	0	0	656		121	121	216
		Las Palmas	87	5.438	0	87	0	87		16	16	16
		Las Pampas	1,117	5.422	0	1,117	0	1,117		206	206	253
		Mayfair	1,411	4.719	1,411	0	0	1,411		299	299	341
		McAdoo Acres	493	5.418	0	493	0	493		91	91	108
		Melton Place	211	5.410	0	211	0	211		39	39	78
		Mesa Verde	179	5.424	179	0	0	179		33	33	52
		Mission Trail Estate	1,572	5.421	0	1,572	0	1,572		290	290	546
		Mobile Haven Estates	448	4.716	448	0	0	448		95	95	142
		Morning Glory Manor	580	5.421	580	0	0	580		107	107	144
		Nuway	245	4.712	245	0	0	245		52	52	17
		Paso del Rey	23	4.600	0	23	0	23		5	5	5

Continued

Table A.8 (continued)
El Paso County Project Summary

Project Status	Project Name / Sponsor	Colonia Name	# of Residents	Density	# Svd. by Public Water	# Not Svd. by Public Water	# Svd. by Central Waste-water	# Not Svd. by Central Waste-water	# Svd. by Project	# of Dwellings	# of Occupied Lots	Total # of Lots
Colonias Not Served in EDAP (continued)		Paso View North	171	4.500	171	0	0	171		38	38	38
		Polkinghorn	160	4.706	160	0	0	160		34	34	60
		Ponderosa MHP	755	4.719	755	0	0	755		160	160	161
		Prado Verde	326	4.725	326	0	0	326		69	69	117
		Sanchez	119	5.409	0	119	0	119		22	22	22
		Schuman Est.	227	4.729	227	0	0	227		48	48	64
		Serene Acres	33	4.714	33	0	0	33		7	7	9
		Sierra Meadows	22	5.500	22	0	0	22		4	4	16
		Socorro Mission	846	5.423	846	0	0	846		156	156	191
		Sparks	3,060	5.240	3,060	0	0	3,060		584	584	1,598
		Sunshine Acres	87	5.438	87	0	0	87		16	16	17
		Tract 27-3D/4	87	5.438	0	87	0	87		16	16	16
		Varela	157	7.850	157	0	0	157		20	20	29
		Vista Larga	60	5.455	0	60	0	60		11	11	11
		Warren Allen Road	260	5.417	0	260	0	260		48	48	48
		Wilbourn	11	5.500	0	11	0	11		2	2	24
		Wildhorse Valle	168	5.419	168	0	0	168		31	31	34
NOT SERVED BY EDAP TOTAL - EL PASO COUNTY		**Number of Colonias: 57**	**26,380**	**5.222**	**14,466**	**11,914**	**0**	**26,380**	**0**	**5,052**	**5,048**	**8,159**

Table A.9
Frio County Project Summary

Project Status	Project Name / Sponsor	Colonia Name	# of Residents	Density	# Svd. by Public Water	# Not Svd. by Public Water	# Svd. by Central Waste-water	# Not Svd. by Central Waste-water	# Svd. by Project	# of Dwellings	# of Occupied Lots	Total # of Lots
Colonias Not Served in EDAP												
		Alta Vista & Frio Heights	450	4.500	450	0	113	337		100	100	185
		Big Foot	105	3.500	105	0	0	105		30	30	200
		Moore	175	3.500	175	0	0	175		50	50	80
NOT SERVED BY EDAP - TOTAL FRIO COUNTY		**Number of Colonias: 3**	**730**	**4.056**	**730**	**0**	**113**	**617**	**0**	**180**	**180**	**465**

Table A.10
Hidalgo County Project Summary

Project Status	Project Name / Sponsor	Colonia Name	# of Residents	Density	# Svd. by Public Water	# Not Svd. By Public Water	# Svd. By Central Waste-water	# Not Svd. By Central Waste-water	# Svd. By Project	# of Dwellings	# of Occupied Lots	Total # of Lots
Completed												
	City of Edinburg (Lull)											
		Lull	1,296	4.500	1,296	0	1296	0	1,296	288	288	348
	TOTAL - City of Edinburg (Lull)		**1,296**	**4.500**	**1,296**	**0**	**1296**	**0**	**1,296**	**288**	**288**	**348**
Under Construction												
	City of Edinburg (Faysville)											
		281 Estates	108	4.500	108	0	0	108		24	24	44
		Americana	135	4.500	135	0	0	135		30	30	50
		Bar #2	144	4.500	144	0	0	144		32	32	60
		Bar #5	459	4.500	459	0	0	459		102	102	137
		Doolittle Acres	23	4.600	23	0	0	23		5	5	9
		Edinburg Acres	54	3.857	54	0	0	54		14	14	18
		El Seco	99	4.500	99	0	0	99		22	22	27
		Faysville	320	4.507	320	0	0	320		71	71	71
		Fenton Minnie E	54	4.500	54	0	0	54		12	12	14
		Hern Subdivision	23	4.600	23	0	0	23		5	5	5
		Highway Frontage	54	4.500	54	0	0	54		12	12	20
		Hilda #1	171	4.500	171	0	0	171		38	38	43
		Lake Citrus Est	63	4.500	63	0	0	63		14	14	36
		Meadow Lands	140	4.516	140	0	0	140		31	31	40
		Merrill	59	4.538	59	0	0	59		13	13	13
		Monte Cristo	72	4.500	72	0	0	72		16	16	14
		Montemayor	36	4.500	36	0	0	36		8	8	8
		Pralle	72	4.500	72	0	0	72		16	16	14
		Puerta Del Sol #1	86	4.526	86	0	0	86		19	19	38
		Sandy Ridge	293	4.508	293	0	0	293		65	65	65
		Santa Cruz Est	99	4.500	99	0	0	99		22	22	29
		Santa Cruz Orange Gardens	135	4.500	135	0	0	135		30	30	38

Continued

Table A.10 (continued)
Hidalgo County Project Summary

Project Status	Project Name / Sponsor	Colonia Name	# of Residents	Density	# Svd. by Public Water	# Not Svd. By Public Water	# Svd. By Central Waste-water	# Not Svd. By Central Waste-water	# Svd. By Project	# of Dwellings	# of Occupied Lots	Total # of Lots
Under Construction (continued)		Sauceda Subdivision	18	4.500	18	0	0	18		4	4	4
		Spring Green Subdivision	68	4.533	68	0	0	68		15	15	17
		Summerwood Subd	117	4.500	117	0	0	117		26	26	36
		Triple C	144	4.500	144	0	0	144		32	32	32
		Twin Lake Subd	95	4.524	95	0	0	95		21	21	21
	TOTAL - City of Edinburg (Faysville)		**3,141**	**4.494**	**3,141**	**0**	**0**	**3,141**	**3,141**	**699**	**699**	**903**
	City of Mission (Granjeno & Madero)											
		Granjeno	488	5.880	488	0	0	488		83	83	65
		Madero	873	6.820	873	0	0	873		128	390	447
	TOTAL - City of Mission (Granjeno & Madero)		**1,361**	**6.450**	**1,361**	**0**	**0**	**1,361**	**1,361**	**211**	**473**	**512**
Plans and Specifications												
	City of Alton											
		Aloha Village	113	4.520	113	0	0	113		25	25	149
		Amberland	36	4.500	36	0	0	36		8	8	14
		Armstrong Alton	140	4.516	140	0	0	140		31	31	38
		Bella Vista Subd	50	4.545	50	0	0	50		11	11	34
		Bondine-Birdwell #1	54	4.500	54	0	0	54		12	12	20
		Bryan Acres	59	4.538	59	0	0	59		13	13	16
		Cantu, Jose #1	54	4.500	54	0	0	54		12	12	16
		Corina's Corner	23	4.600	23	0	0	23		5	5	8
		de la Garza Subdivision	135	4.500	135	0	0	135		30	30	32
		Dude Hill #1	59	4.538	59	0	0	59		13	13	22
		Dude Hill #2	27	4.500	27	0	0	27		6	6	10
		Friendly Acres	95	4.524	95	0	0	95		21	21	35
		Glasscock Est	45	4.500	45	0	0	45		10	10	16
		Glenshire Est.	32	4.571	32	0	0	32		7	7	14

Continued

Table A.10 (continued)
Hidalgo County Project Summary

Project Status	Project Name / Sponsor	Colonia Name	# of Residents	Density	# Svd. by Public Water	# Not Svd. By Public Water	# Svd. By Central Waste-water	# Not Svd. By Central Waste-water	# Svd. By Project	# of Dwellings	# of Occupied Lots	Total # of Lots
Plans and Specifications (continued)		Inspiration (Unrecorded)	149	4.515	149	0	0	149		33	33	40
		Inspiration Heights	27	4.500	27	0	0	27		6	6	11
		Jardin Terrace	81	4.500	81	0	0	81		18	18	30
		North Cross Est	131	4.517	131	0	0	131		29	29	27
		Nuevo Alton	59	4.538	59	0	0	59		13	13	22
		Nuevo Alton Sub #1	1,121	4.502	1,121	0	0	1,121		249	249	383
		Palm Acres #1	23	4.600	23	0	0	23		5	5	9
		Palm Heights Subd	221	4.510	221	0	0	221		49	49	49
		Palm Lake #1 thru 4	2,579	4.501	2,579	0	0	2,579		573	573	955
		Rabbit Patch #1 & 2	131	4.517	131	0	0	131		29	29	49
		Rancho Chapparral	99	4.500	99	0	0	99		22	22	26
		Rancho Grande	36	4.500	36	0	0	36		8	8	66
		Salas Subd	72	4.500	72	0	0	72		16	16	20
		Stewart Palms	54	4.500	54	0	0	54		12	12	20
		Stewart Place	32	4.571	32	0	0	32		7	7	60
		Stewart Place Comm	27	4.500	27	0	0	27		6	6	10
		Stewart South	117	4.500	117	0	0	117		26	26	44
		Tierra Est. #1	302	4.507	302	0	0	302		67	67	111
		Tierra Estates #2	72	4.500	72	0	0	72		16	16	48
		Tres Amigos Subdivision	23	4.600	23	0	0	23		5	5	6
		Trevino Subdivision	90	4.500	90	0	0	90		20	20	32
		Tri-City #1	239	4.509	239	0	0	239		53	53	99
		Tri-City #2	383	4.506	383	0	0	383		85	85	147
		Val Verde	108	4.500	108	0	0	108		24	14	14
		Val Verde Grove	23	4.600	23	0	0	23		5	5	20
		Vereda Tropical	81	4.500	81	0	0	81		18	18	20
	TOTAL - City of Alton		**7,202**	**4.507**	**7,202**	**0**	**0**	**7,202**	**7,202**	**1,598**	**1,588**	**2,742**
	City of Pharr (Las Milpas)											
		Beamsley	360	4.500	360	0	0	360		80	80	80

Continued

Table A.10 (continued)
Hidalgo County Project Summary

Project Status	Project Name / Sponsor	Colonia Name	# of Residents	Density	# Svd. by Public Water	# Not Svd. By Public Water	# Svd. By Central Waste-water	# Not Svd. By Central Waste-water	# Svd. By Project	# of Dwellings	# of Occupied Lots	Total # of Lots
Plans and Specifications (continued)		Colonia Esperanza	458	4.771	458	0	0	458		96	96	96
		El Sol	106	5.048	106	0	0	106		21	21	21
		Hidalgo Park	2,657	5.061	2,657	0	0	2,657		525	525	525
		La Quinta	420	4.828	420	0	0	420		87	87	87
		Las Brisas	292	5.034	292	0	0	292		58	58	58
		Las Fuentas	81	5.400	81	0	0	81		15	15	15
		Las Haciendas	352	5.176	352	0	0	352		68	68	68
		Las Milpas	627	4.786	627	0	0	627		131	131	131
		Las Milpas Road	620	5.254	620	0	0	620		118	118	118
		Los Ranchitos	665	5.278	665	0	0	665		126	126	126
		Lot 113	135	4.500	135	0	0	135		30	30	30
		Moore Road	48	4.800	48	0	0	48		10	10	15
		Palo Verde	327	6.412	327	0	0	327		51	51	51
		Ridge Road	149	4.515	149	0	0	149		33	33	33
		Southfork Estates	321	3.821	321	0	0	321		84	84	84
		Thrasher Terrace	81	4.765	81	0	0	81		17	17	17
		Universal Estates	204	5.368	204	0	0	204		38	38	38
		Villas del Valle	670	5.492	670	0	0	670		122	122	122
		Yvonne	16	3.200	16	0	0	16		5	5	5
	TOTAL - City of Pharr (Las Milpas)		**8,589**	**5.008**	**8,589**	**0**	**0**	**8,589**	**8,589**	**1,715**	**1,715**	**1,720**
	City of Weslaco											
		Agua Dulce	319	4.431	319	0	0	319		72	72	96
		Angela	253	5.060	0	253	0	253		50	50	36
		Babb RC #1,2 & 3	696	4.350	696	0	0	696		160	160	243
		Bellaire	96	6.000	96	0	0	96		16	16	124
		Cuellar #1,2,3 & 4	482	4.050	482	0	0	482		119	119	167
		Llano Grande	662	4.904	662	0	0	662		135	135	179
		Los Castillos Est.	133	5.320	133	0	0	133		25	25	33

Continued

Table A.10 (continued)
Hidalgo County Project Summary

Project Status	Project Name / Sponsor	Colonia Name	# of Residents	Density	# Svd. by Public Water	# Not Svd. By Public Water	# Svd. By Central Waste-water	# Not Svd. By Central Waste-water	# Svd. By Project	# of Dwellings	# of Occupied Lots	Total # of Lots
Plans and Specifications (continued)		Rockwell (Ramosville)	216	3.857	216	0	0	216		56	37	56
	TOTAL - City of Weslaco		**2,857**	**4.513**	**2,604**	**253**	**0**	**2,857**	**2,857**	**633**	**614**	**934**
	Hidalgo County (deAnda & Saenz)											
		DeAnda & Saenz	132	4.400	0	132	0	132	132	30	30	35
	TOTAL - Hidalgo County (deAnda & Saenz)		**132**	**4.400**	**0**	**132**	**0**	**132**	**132**	**30**	**30**	**35**
Facilities Planning												
	Alamo											
		South Tower Estates	1,425	5.144	1,425	0	0	1,425	1,425	277	277	277
	TOTAL - Alamo		**1,425**	**5.144**	**1,425**	**0**	**0**	**1,425**	**1,425**	**277**	**277**	**277**
	City of Edinburg (Northwest)											
		Alma	99	4.500	99	0	0	99		22	22	36
		Alvacan	126	4.500	126	0	0	126		28	28	46
		Austin Gardens	59	4.538	59	0	0	59		13	13	22
		Bar 6 Subd.	302	4.507	302	0	0	302		67	67	91
		Batson Gardens	32	4.571	32	0	0	32		7	7	12
		Beretta Subdivision	36	4.500	36	0	0	36		8	8	12
		Crouse	27	4.500	27	0	0	27		6	6	10
		Floresta	54	4.500	54	0	0	54		12	12	20
		Hacienda Del Bronco #1 & 2	230	4.510	230	0	0	230		51	51	60
		Hoehn Drive	441	4.500	441	0	0	441		98	98	164
		Hoehn Rd Estates	63	4.500	63	0	0	63		14	14	24
		MAX	86	4.526	86	0	0	86		19	19	31
		McColl Est #1	149	4.515	149	0	0	149		33	33	55
		Milyca	45	4.500	45	0	0	45		10	10	16
		Monte Cristo Acres	234	4.500	234	0	0	234		52	52	86
		North McColl	36	4.500	36	0	0	36		8	8	14
		R.C.W.	266	4.508	266	0	0	266		59	59	98

Continued

Table A.10 (continued)
Hidalgo County Project Summary

Project Status	Project Name / Sponsor	Colonia Name	# of Residents	Density	# Svd. by Public Water	# Not Svd. By Public Water	# Svd. By Central Waste-water	# Not Svd. By Central Waste-water	# Svd. By Project	# of Dwellings	# of Occupied Lots	Total # of Lots
Facilities Planning (continued)		River Bend	77	4.529	77	0	0	77		17	17	29
		Rodgers Lake Estates	45	4.500	45	0	0	45		10	10	19
		Tagle, Robert	32	4.571	32	0	0	32		7	7	11
		Tierra #1	302	4.507	302	0	0	302		67	67	111
		West Haven	36	4.500	36	0	0	36		8	8	14
	TOTAL - City of Edinburg (Northwest)		**2,777**	**4.508**	**2,777**	**0**	**0**	**2,777**	**2,777**	**616**	**616**	**981**
	Donna											
		Avila I B	63	4.500	63	0	0	63		14	14	14
		Balli #2	207	4.500	207	0	0	207		46	46	77
		Balli Est	437	4.505	437	0	0	437		97	97	162
		Benita	108	4.500	108	0	0	108		24	22	42
		Clark	113	4.520	113	0	0	113		25	25	41
		Koenig Winter Resort	392	4.506	392	0	0	392		87	87	87
		Lunar Heights	405	4.500	405	0	0	405		90	90	170
		Remuda RV Park	90	4.500	90	0	0	90		20	20	24
		River Road	167	4.514	167	0	0	167		37	37	62
		Schroeder	608	4.504	608	0	0	608		135	135	225
		South Donna Subd.	306	5.464	306	0	0	306		56	56	56
		Sun Groves Park	167	4.514	167	0	0	167		37	37	49
		Tierra Bella	122	4.519	122	0	0	122		27	27	36
		Tierra del Sol Est	27	4.500	27	0	0	27		6	6	10
		Tierra Prieta	212	4.511	212	0	0	212		47	47	78
		Val Verde North	270	4.500	270	0	0	270		60	60	100
		Villa Donna	405	4.500	405	0	0	405		90	90	94
		Village Groves #2	338	4.507	338	0	0	338		75	75	94
	TOTAL - Donna		**4,437**	**4.560**	**4,437**	**0**	**0**	**4,437**	**4,437**	**973**	**971**	**1,421**
	Elsa											
		Bar #3	216	4.500	216	0	0	216		48	48	98
		Bernal Heights #1	104	4.522	104	0	0	104		23	23	21
		Bernal Heights #2	32	4.571	32	0	0	32		7	7	12

Continued

Table A.10 (continued)
Hidalgo County Project Summary

Project Status	Project Name / Sponsor	Colonia Name	# of Residents	Density	# Svd. by Public Water	# Not Svd. By Public Water	# Svd. By Central Waste-water	# Not Svd. By Central Waste-water	# Svd. By Project	# of Dwellings	# of Occupied Lots	Total # of Lots
Facilities Planning (continued)		Brenda Gay	45	4.500	45	0	0	45		10	10	12
		Cana De Azuear	140	4.516	140	0	0	140		31	31	44
		Casas Del Valle	180	4.500	180	0	0	180		40	40	60
		El Mesquite	81	4.500	81	0	0	81		18	18	20
		El Mesquite 1	135	4.500	135	0	0	135		30	30	48
		El Monte	90	4.500	90	0	0	90		20	20	31
		Engleman Est.	234	4.500	234	0	0	234		52	52	86
		Garza #1	32	4.571	32	0	0	32		7	7	12
		Garza Subd (Phase 2)	144	4.500	144	0	0	144		32	32	52
		George Lookingbill 1	95	4.524	95	0	0	95		21	21	28
		George Lookingbill 2	63	4.500	63	0	0	63		14	14	23
		Hacienda De Los Vega	54	4.500	54	0	0	54		12	12	136
		Hoyt J T	23	4.600	23	0	0	23		5	5	9
		L & P	59	4.538	59	0	0	59		13	13	20
		Mile 16	99	4.500	99	0	0	99		22	22	44
		Oak	45	4.500	45	0	0	45		10	10	14
		Rick Sub.	27	4.500	27	0	0	27		6	6	14
		Salinas-Hinojosa	176	4.513	176	0	0	176		39	39	100
		Ten Acres	95	4.524	95	0	0	95		21	21	24
	TOTAL - Elsa		**2,169**	**4.509**	**2,169**	**0**	**0**	**2,169**	**2,169**	**481**	**481**	**908**
	HC - Urban Regional WW Planning											
		Adkins	113	4.520	113	0	0	113		25	25	42
		Alamo Orchards	126	4.500	126	0	0	126		28	28	46
		Alberta	72	4.500	72	0	0	72		16	16	30
		Alberta Acres	59	4.538	59	0	0	59		13	13	21
		Alberta Est. #2	69	4.600	69	0	0	69		15	15	59
		Albino Est.	45	4.500	45	0	0	45		10	10	32
		Alvarez	45	4.500	0	45	0	45		10	10	9
		Arriaga	14	4.667	14	0	0	14		3	3	8
		Arroyo Park	72	4.500	72	0	0	72		16	16	26

Continued

Table A.10 (continued)
Hidalgo County Project Summary

Project Status	Project Name / Sponsor	Colonia Name	# of Residents	Density	# Svd. by Public Water	# Not Svd. By Public Water	# Svd. By Central Waste-water	# Not Svd. By Central Waste-water	# Svd. By Project	# of Dwellings	# of Occupied Lots	Total # of Lots
Facilities Planning (continued)		Ash Country	81	4.500	81	0	0	81		18	18	36
		Bar #7	230	4.510	230	0	0	230		51	51	75
		Bibleville Trailer Park	216	4.500	216	0	0	216		48	48	80
		Big John	18	4.500	18	0	0	18		4	4	26
		Brown Acres	95	4.524	95	0	0	95		21	21	18
		Carroll Road Acres	32	4.571	32	0	0	32		7	7	10
		Chapa	166	4.611	0	166	0	166		36	36	34
		Chapa #3	230	4.510	230	0	0	230		51	51	88
		Citrus Ranchitos	32	4.571	32	0	0	32		7	7	12
		Closner Sub.	221	4.510	221	0	0	221		49	49	57
		Cole	117	4.500	117	0	0	117		26	26	44
		Colonia Boyes	90	4.500	0	90	0	90		20	20	26
		Colonia Esperanza	Already served under City of Pharr (Las Milpas).									
		Colonia Esperanza #3	41	4.556	41	0	0	41		9	9	15
		Colonia Guadalupe	68	4.533	68	0	0	68		15	16	26
		Colonia Tijerina	45	4.500	45	0	0	45		10	10	15
		Colonia Tijerina	126	4.500	0	126	0	126		28	28	46
		Colonia Whalen Rd	50	4.545	50	0	0	50		11	11	15
		Country Living Est #2	45	4.500	45	0	0	45		10	10	10
		Country Terrace Est	36	4.500	36	0	0	36		8	8	13
		Country Village #1	113	4.520	113	0	0	113		25	25	59
		Country Village #2	252	4.500	252	0	0	252		56	56	76
		D. T. Villareal	77	4.529	77	0	0	77		17	17	31
		Dimas	36	4.500	36	0	0	36		8	8	12
		Donna Heights North	189	4.500	189	0	0	189		42	42	71
		Eagle Heights	41	4.556	41	0	0	41		9	9	15
		Ebony Acres	63	4.500	63	0	0	63		14	14	24
		El Charro #2	329	4.507	329	0	0	329		73	73	97
		Esperanza Est	54	4.500	54	0	0	54		12	12	20
		Evergreen Est.	72	4.500	72	0	0	72		16	16	28

Continued

Table A.10 (continued)
Hidalgo County Project Summary

Project Status	Project Name / Sponsor	Colonia Name	# of Residents	Density	# Svd. by Public Water	# Not Svd. By Public Water	# Svd. By Central Waste-water	# Not Svd. By Central Waste-water	# Svd. By Project	# of Dwellings	# of Occupied Lots	Total # of Lots
Facilities Planning (continued)		Garden of Christus	23	4.600	23	0	0	23		5	5	8
		Green Valley Acres	95	4.524	95	0	0	95		21	21	34
		Hacienda De Los Vega	Already served under Elsa..									
		Hi-Land	45	4.500	45	0	0	45		10	10	16
		High Chapparral	135	4.500	135	0	0	135		30	13	16
		Highland Memorial Park	45	4.500	45	0	0	45		10	10	16
		Isaacs	189	4.500	189	0	0	189		42	42	70
		Jacksons New World Dev #1	59	4.538	59	0	0	59		13	13	18
		Jacksons New World Dev #2	27	4.500	27	0	0	27		6	6	6
		L J #1	176	4.513	176	0	0	176		39	39	74
		La Flor Est.	90	4.500	90	0	0	90		20	20	20
		La Flor Gardens	72	4.500	72	0	0	72		16	16	20
		La Paloma #1 & 2	59	4.538	59	0	0	59		13	13	28
		La Quinta Est.	117	4.500	117	0	0	117		26	26	43
		LaBlanca Heights	122	4.519	122	0	0	122		27	27	38
		Laborsita	32	4.571	0	32	0	32		7	7	12
		Las Brisas Est	450	4.500	450	0	0	450		100	100	150
		Las Villas Del Valle	86	4.526	86	0	0	86		19	19	32
		Los Encinos #1	104	4.522	104	0	0	104		23	23	38
		Los Encinos #2	27	4.500	27	0	0	27		6	6	10
		Los Tinacos	41	4.556	41	0	0	41		9	9	20
		Matt	99	4.500	0	99	0	99		22	22	51
		Mesquite Acres	153	4.500	153	0	0	153		34	34	48
		Midway Manor	108	4.500	108	0	0	108		24	24	48
		Mile 12 N Subd.	126	4.500	126	0	0	126		28	28	28
		Mile Doce West	86	4.526	86	0	0	86		19	19	27
		Minnesota Acres	27	4.500	27	0	0	27		6	6	10

Continued

Table A.10 (continued)
Hidalgo County Project Summary

Project Status	Project Name / Sponsor	Colonia Name	# of Residents	Density	# Svd. by Public Water	# Not Svd. By Public Water	# Svd. By Central Waste-water	# Not Svd. By Central Waste-water	# Svd. By Project	# of Dwellings	# of Occupied Lots	Total # of Lots
Facilities Planning (continued)		Moore Road Subd	32	4.571	32	0	0	32		7	7	12
		Morningside Est	131	4.517	131	0	0	131		29	29	49
		Morningside South	86	4.526	86	0	0	86		19	19	28
		Muniz	531	4.500	531	0	0	531		118	118	186
		Murillo	23	4.600	23	0	0	23		5	5	8
		North Alamo Terrace	126	4.500	126	0	0	126		28	28	66
		North Alamo Village	117	4.500	117	0	0	117		26	26	32
		Northside Village #1	275	4.508	275	0	0	275		61	61	98
		Northside Village #2	131	4.517	131	0	0	131		29	29	48
		Old Alamo	45	4.500	45	0	0	45		10	10	3
		Olivarez	135	4.500	135	0	0	135		30	30	34
		Olivarez #10	140	4.516	0	140	0	140		31	31	31
		Olivarez #4	72	4.500	0	72	0	72		16	16	27
		Olivarez #6	27	4.500	0	27	0	27		6	6	10
		Olivarez #8	81	4.500	0	81	0	81		18	18	18
		Owassa Est.	113	4.520	113	0	0	113		25	25	74
		Palma #2	117	4.500	117	0	0	117		26	26	43
		Patio Villas	414	4.500	414	0	0	414		92	92	154
		Plumosa Village	27	4.500	27	0	0	27		6	6	10
		Puesta del Sol	594	4.500	594	0	0	594		132	132	193
		Quiet Village 2	329	4.507	329	0	0	329		73	73	293
		R & G	77	4.529	77	0	0	77		17	17	19
		Rambo Est.	77	4.529	77	0	0	77		17	17	26
		Rancho Nuevo	257	4.509	257	0	0	257		57	57	95
		Reina del Sol	140	4.516	140	0	0	140		31	31	76
		Road Runner #2	113	4.520	113	0	0	113		25	25	42
		Rosalito	27	4.500	27	0	0	27		6	6	14
		Royal Palms Mobil Hm Comm	140	4.516	140	0	0	140		31	31	51
		Ruthuen #1	77	4.529	77	0	0	77		17	17	17
		Ruthven #2	90	4.500	90	0	0	90		20	20	34

Continued

Table A.10 (continued)
Hidalgo County Project Summary

Project Status	Project Name / Sponsor	Colonia Name	# of Residents	Density	# Svd. by Public Water	# Not Svd. By Public Water	# Svd. By Central Waste-water	# Not Svd. By Central Waste-water	# Svd. By Project	# of Dwellings	# of Occupied Lots	Total # of Lots
Facilities Planning (continued)		Serendipity Way	45	4.500	45	0	0	45		10	10	16
		Sherry	32	4.571	32	0	0	32		7	7	10
		Sierra Terrace	32	4.571	32	0	0	32		7	7	12
		Southport	68	4.533	68	0	0	68		15	15	20
		Southridge Park	23	4.600	23	0	0	23		5	5	7
		Southside Village	176	4.513	176	0	0	176		39	39	65
		Sundowners Retire	162	4.500	162	0	0	162		36	36	44
		Sunny Brook	140	4.516	140	0	0	140		31	31	42
		Sunrise	410	4.505	410	0	0	410		91	91	116
		Sunrise #2	743	4.503	743	0	0	743		165	165	237
		Sunrise Hill	599	4.504	599	0	0	599		133	133	222
		Terry Subd.	153	4.500	153	0	0	153		34	34	37
		Tessa Sub.	27	4.500	27	0	0	27		6	6	8
		The Highlands	32	4.571	32	0	0	32		7	7	11
		The Stables	36	4.500	0	36	0	36		8	8	14
		Thirty-Six Palms Terrace	63	4.500	63	0	0	63		14	14	60
		Thompson Road	122	4.519	122	0	0	122		27	27	38
		Tierra Bonita 1 & 2	261	4.500	261	0	0	261		58	58	99
		Tierra Del Valle	86	4.526	86	0	0	86		19	19	35
		Tony Subdivision	32	4.571	32	0	0	32		7	7	15
		Tower Heights	50	4.545	50	0	0	50		11	11	24
		Tower Sub.	113	4.520	113	0	0	113		25	25	21
		Towne East	113	4.520	113	0	0	113		25	25	30
		Trenton Acres	36	4.500	36	0	0	36		8	8	12
		Trenton Manor	194	4.512	194	0	0	194		43	43	72
		Trenton Terrace	68	4.533	68	0	0	68		15	15	25
		Trophy Park	1,436	4.502	1,436	0	0	1,436		319	319	531
		Tropical Farms	185	4.512	185	0	0	185		41	41	71
		Val Verde North	90	4.500	90	0	0	90		20	20	20

Continued

Table A.10 (continued)
Hidalgo County Project Summary

Project Status	Project Name / Sponsor	Colonia Name	# of Residents	Density	# Svd. by Public Water	# Not Svd. By Public Water	# Svd. By Central Waste-water	# Not Svd. By Central Waste-water	# Svd. By Project	# of Dwellings	# of Occupied Lots	Total # of Lots
Facilities Planning (continued)		Val Verde Park	27	4.500	27	0	0	27		6	6	6
		Valley Star Acres	63	4.500	0	63	0	63		14	14	23
		Veronica Ann	63	4.500	63	0	0	63		14	14	23
		Victoria Acres #1	45	4.500	45	0	0	45		10	10	10
		Villa D Val	86	4.526	86	0	0	86		19	19	32
		Villa Del Mundo	261	4.500	261	0	0	261		58	58	94
		Villa Del Sol	239	4.509	239	0	0	239		53	53	82
		Walston Farms	275	4.508	275	0	0	275		61	61	114
		Welch	50	4.545	50	0	0	50		11	11	18
		Whitewing Subdivision	167	4.514	167	0	0	167		37	37	62
	TOTAL - HC - Urban Regional WW Planning		**17,720**	**4.510**	**16,743**	**977**	**0**	**17,720**	**17,720**	**3,929**	**3,913**	**6,402**
	Hidaldo County (Sanchez Ranch)											
		Sanchez Ranch	550	5.978	550	0	0	550	550	92	92	100
	TOTAL - Hidaldo County (Sanchez Ranch)		**550**	**5.978**	**550**	**0**	**0**	**550**	**550**	**92**	**92**	**100**
	La Joya WSC											
		Benavides #1	581	4.504	581	0	0	581		129	129	131
		Benavides Subd #2	374	4.506	374	0	0	374		83	83	69
		Blue Star Enterprises #2	122	4.519	122	0	0	122		27	27	67
		Citrus Lake Est	153	4.500	153	0	0	153		34	34	106
		Colonia Rafael	144	4.500	144	0	0	144		32	32	52
		Colonia San Miguel	113	4.520	113	0	0	113		25	25	45
		Cuevitas	50	4.545	50	0	0	50		11	11	11
		El Flaco, El Flaco Chiquito	342	4.500	342	0	0	342		76	76	127
		Fisher	306	4.500	306	0	0	306		68	68	154
		Flores	167	4.514	167	0	0	167		37	37	40
		Havana	99	4.500	99	0	0	99		22	22	22

Continued

Table A.10 (continued)
Hidalgo County Project Summary

Project Status	Project Name / Sponsor	Colonia Name	# of Residents	Density	# Svd. by Public Water	# Not Svd. By Public Water	# Svd. By Central Waste-water	# Not Svd. By Central Waste-water	# Svd. By Project	# of Dwellings	# of Occupied Lots	Total # of Lots
Facilities Planning (continued)		Havana Lomas	77	4.529	77	0	0	77		17	17	29
		Havana Lomas #4	117	4.500	117	0	0	117		26	26	43
		La Aurora	95	4.524	95	0	0	95		21	21	28
		La Hermosa	77	4.529	77	0	0	77		17	17	29
		Las Cuevas #2	342	4.500	342	0	0	342		76	76	269
		Los Ebanos	134	5.583	134	0	0	134		24	24	52
		Rancho	135	4.500	135	0	0	135		30	30	45
		Rodrique/Sullivan City	95	4.524	95	0	0	95		21	21	26
		South Fork	284	4.508	284	0	0	284		63	63	84
		Vales Subd	113	4.520	113	0	0	113		25	25	37
		Valle Vista	198	4.500	198	0	0	198		44	44	103
		Villa Est	63	4.500	63	0	0	63		14	14	68
		Western Est. #1 & 2	131	4.517	131	0	0	131		29	29	34
	TOTAL - La Joya WSC		**4,312**	**4.534**	**4,312**	**0**	**0**	**4,312**	**4,312**	**951**	**951**	**1,671**
	McAllen											
		Acosta	59	4.538	59	0	0	59		13	13	42
		Acosta 107	50	4.545	50	0	0	50		11	11	39
		Annalisa Subd.	32	4.571	32	0	0	32		7	7	12
		Basham #14	153	4.500	153	0	0	153		34	34	52
		Bougainvillea	86	4.526	86	0	0	86		19	19	31
		Brandon Lake Subdivision	23	4.600	23	0	0	23		5	5	16
		Calma Estates #1	18	4.500	18	0	0	18		4	4	12
		Calma Estates #2	41	4.556	41	0	0	41		9	9	15
		Calma Estates #3	27	4.500	27	0	0	27		6	6	14
		Citrus City Lake #1	113	4.520	113	0	0	113		25	25	30
		Country Colony	36	4.500	36	0	0	36		8	8	14
		Curl Tex	32	4.571	32	0	0	32		7	7	12
		Devan Estates	99	4.500	99	0	0	99		22	22	36
		Diaz Subdivision	126	4.500	126	0	0	126		28	28	30

Continued

Table A.10 (continued)
Hidalgo County Project Summary

Project Status	Project Name / Sponsor	Colonia Name	# of Residents	Density	# Svd. by Public Water	# Not Svd. By Public Water	# Svd. By Central Waste-water	# Not Svd. By Central Waste-water	# Svd. By Project	# of Dwellings	# of Occupied Lots	Total # of Lots
Facilities Planning (continued)		Garza LR	99	4.500	99	0	0	99		22	22	37
		Gort Peter Subdivision	45	4.500	45	0	0	45		10	10	10
		Granada Est	59	4.538	59	0	0	59		13	13	38
		Haven Subdivision	54	4.500	54	0	0	54		12	12	20
		Los Padres	90	4.500	90	0	0	90		20	20	34
		Los Terrazos	27	4.500	27	0	0	27		6	6	24
		Monte Alban	72	4.500	72	0	0	72		16	16	26
		Monte Cristo	Already served under City of Edinburg (Faysville).									
		Newkirk Subdivision	63	4.500	63	0	0	63		14	14	14
		Palmeras	54	4.500	54	0	0	54		12	12	20
		Plantation Oaks	45	4.500	45	0	0	45		10	10	20
		Ramon Leal	86	4.526	86	0	0	86		19	19	16
		Regency Acres #2	63	4.500	63	0	0	63		14	14	24
		Rena Rae	180	4.500	180	0	0	180		40	40	45
		RLDS (Amistad)	221	4.510	221	0	0	221		49	49	40
		Shary Country Acres	32	4.571	32	0	0	32		7	7	17
		Shary Groves Est.	86	4.526	86	0	0	86		19	19	12
		Simpatico Acres	54	4.500	54	0	0	54		12	12	20
		Timberhill Villa #4	135	4.500	135	0	0	135		30	30	40
		Twin Roads Subd	104	4.522	104	0	0	104		23	21	35
		Upper Sharyland	27	4.500	27	0	0	27		6	6	22
		Ware Colony Subdiv.	27	4.500	27	0	0	27		6	6	14
		Ware Country Subd 1 & 2	32	4.571	32	0	0	32		7	7	31
		Ware Del Norte	41	4.556	41	0	0	41		9	9	14
		Ware Estates	27	4.500	27	0	0	27		6	6	11
		Ware Shadows	99	4.500	99	0	0	99		22	22	48
		Ware West Subdivision	27	4.500	27	0	0	27		6	6	13
	TOTAL - McAllen		**2,744**	**4.513**	**2,744**	**0**	**0**	**2,744**	**2,744**	**608**	**606**	**1,000**

Continued

Table A.10 (continued)
Hidalgo County Project Summary

Project Status	Project Name / Sponsor	Colonia Name	# of Residents	Density	# Svd. by Public Water	# Not Svd. By Public Water	# Svd. By Central Waste-water	# Not Svd. By Central Waste-water	# Svd. By Project	# of Dwellings	# of Occupied Lots	Total # of Lots
	Mercedes											
		Capisallo Park	795	4.848	795	0	0	795		164	164	183
		Chapa	249	3.952	0	249	0	249		63	63	74
		Conner	103	4.292	103	0	0	103		24	24	23
		Eastland Park	262	5.137	262	0	0	262		51	51	51
		Elizabeth & V & C	256	5.565	0	256	0	256		46	46	41
		Heidelberg	506	4.559	506	0	0	506		111	111	207
		Highland	455	4.505	455	0	0	455		101	101	169
		La Mesa	588	5.765	588	0	0	588		102	102	167
		La Milpa	27	4.500	0	27	0	27		6	6	7
		Lorenzana	60	3.333	0	60	0	60		18	18	28
		Los Cerritos	187	3.169	187	0	0	187		59	59	33
		Old Rebel Heights 1 & 2	180	4.865	180	0	0	180		37	37	59
		Olympic	133	4.926	133	0	0	133		27	27	21
		Southern Valley Ests.	45	4.500	45	0	0	45		10	10	44
		Sunrise	130	5.000	0	130	0	130		26	26	16
	TOTAL - Mercedes		**3,976**	**4.705**	**3,254**	**722**	**0**	**3,976**	**3,976**	**845**	**845**	**1,123**
	City of Mission (North Mission)											
		Alturas De Azahares	45	4.500	45	0	0	45		10	10	20
		Arco Iris	212	4.511	212	0	0	212		47	47	52
		Ariel Hinojosa Subd.	36	4.500	36	0	0	36		8	8	11
		Basham # 1	140	4.516	140	0	0	140		31	31	40
		Basham # 2	90	4.500	90	0	0	90		20	20	20
		Basham # 4	77	4.529	77	0	0	77		17	17	28
		Basham # 5	86	4.526	86	0	0	86		19	19	20
		Basham # 6	67	4.467	67	0	0	67		15	15	24
		Basham # 7	81	4.500	81	0	0	81		18	18	18
		Basham # 8	108	4.500	108	0	0	108		24	24	38
		Basham #10	113	4.520	113	0	0	113		25	25	42
		Basham #11	117	4.500	117	0	0	117		26	26	44

Continued

Table A.10 (continued)
Hidalgo County Project Summary

Project Status	Project Name / Sponsor	Colonia Name	# of Residents	Density	# Svd. by Public Water	# Not Svd. By Public Water	# Svd. By Central Waste-water	# Not Svd. By Central Waste-water	# Svd. By Project	# of Dwellings	# of Occupied Lots	Total # of Lots
Facilities Planning (continued)		Basham Subd. (M & B)	131	4.517	131	0	0	131		29	29	39
		Bella Vista Subd	Already served under City of Alton.									
		Blue Rock	59	4.538	59	0	0	59		13	13	20
		Carlos Leal Jr Subd #2	54	4.500	54	0	0	54		12	22	50
		Carlos Leal Jr Subd #2	180	4.500	180	0	0	180		40	40	50
		Castaneda Subd.	50	4.545	50	0	0	50		11	11	8
		Cavazos, Alex	104	4.522	104	0	0	104		23	23	30
		Chacon Estates	54	4.500	54	0	0	54		12	11	11
		Chula Vista Acres	149	4.515	149	0	0	149		33	33	36
		Citrus Hills	50	4.545	50	0	0	50		11	11	18
		Coronado	36	4.500	36	0	0	36		8	8	13
		Country Estates West	95	4.524	95	0	0	95		21	20	20
		Del Norte	50	4.545	50	0	0	50		11	11	26
		Diamond L	86	4.526	86	0	0	86		19	19	48
		El Sol #1 & 2	342	4.500	342	0	0	342		76	76	124
		Elida Subd.	23	4.600	23	0	0	23		5	5	19
		Evie Subd.	198	4.500	198	0	0	198		44	44	44
		Friendly Acres	Already served under City of Alton.									
		Glasscock Est	Already served under City of Alton.									
		Good Valley Ranch	104	4.522	104	0	0	104		23	23	50
		Grovewood Est	45	4.500	45	0	0	45		10	10	18
		Hamlet	23	4.600	23	0	0	23		5	5	5
		Hilda #2	95	4.524	95	0	0	95		21	21	44
		Jenna Estates	50	4.545	50	0	0	50		11	11	18
		Jessan	27	4.500	27	0	0	27		6	6	10
		Kristi Estates	140	4.516	140	0	0	140		31	31	53
		La Homa Acres	23	4.600	23	0	0	23		5	5	12

Continued

Table A.10 (continued)
Hidalgo County Project Summary

Project Status	Project Name / Sponsor	Colonia Name	# of Residents	Density	# Svd. by Public Water	# Not Svd. By Public Water	# Svd. By Central Waste-water	# Not Svd. By Central Waste-water	# Svd. By Project	# of Dwellings	# of Occupied Lots	Total # of Lots
Facilities Planning (continued)		La Homa Acres #2	18	4.500	18	0	0	18		4	4	18
		La Homa Grove Estates	279	4.500	279	0	0	279		62	62	86
		La Homa Road North (Amended)	284	4.508	284	0	0	284		63	63	112
		La Palma	59	4.538	59	0	0	59		13	13	26
		La Suena	72	4.500	72	0	0	72		16	16	93
		Linda Vista Est.	689	4.503	689	0	0	689		153	153	193
		M & S Subdivision	54	4.500	54	0	0	54		12	12	26
		Milagro Subd.	32	4.571	32	0	0	32		7	7	10
		Monger, Boyd Subd.	32	4.571	32	0	0	32		7	7	24
		Monica Acres	72	4.500	72	0	0	72		16	16	24
		Moreno	27	4.500	27	0	0	27		6	6	15
		Palm Drive North #1	27	4.500	27	0	0	27		6	6	13
		Palmhurst Est	63	4.500	63	0	0	63		14	14	23
		Perlas De Naranja	72	4.500	72	0	0	72		16	16	20
		Rabbit Patch #1 & 2	Already served under City of Alton.									
		Randolph Barnett No 1	77	4.529	77	0	0	77		17	17	22
		Randy Ley	72	4.500	72	0	0	72		16	16	36
		Regal Estates	54	4.500	54	0	0	54		12	12	43
		Rush Subd.	32	4.571	32	0	0	32		7	7	7
		Schuerbach Acres	63	4.500	63	0	0	63		14	14	22
		Sendero	41	4.556	41	0	0	41		9	9	15
		Stonegate #1 & 2	185	4.512	185	0	0	185		41	41	68
		Storylane	68	4.533	68	0	0	68		15	15	20
		Tangerine Est	63	4.500	63	0	0	63		14	14	36
		Thompson	54	4.500	54	0	0	54		12	12	29
		Vereda Tropical	Already served under City of Alton.									

Continued

Table A.10 (continued)
Hidalgo County Project Summary

Project Status	Project Name / Sponsor	Colonia Name	# of Residents	Density	# Svd. by Public Water	# Not Svd. By Public Water	# Svd. By Central Waste-water	# Not Svd. By Central Waste-water	# Svd. By Project	# of Dwellings	# of Occupied Lots	Total # of Lots
Facilities Planning (continued)		Villa Capri	41	4.556	41	0	0	41		9	9	21
	TOTAL - City of Mission (North Mission)		**5,870**	**4.512**	**5,870**	**0**	**0**	**5,870**	**5,870**	**1,301**	**1,309**	**2,125**
	Palmview											
		Acevedo #3	225	4.500	225	0	0	225		50	50	62
		Acevedo #4	297	4.500	297	0	0	297		66	66	85
		Akin Development Subd	72	4.235	72	0	0	72		17	17	29
		Alta Vista	81	4.500	81	0	0	81		18	18	41
		Alysonders Est	32	4.571	32	0	0	32		7	7	29
		Americana Grove	54	4.500	54	0	0	54		12	12	28
		Americana Grove #2 (Amended)	117	4.500	117	0	0	117		26	26	56
		Arco Iris	Already served under City of Mission (North Mission).									
		Ariel Hinojosa Subd No 3	95	4.524	95	0	0	95		21	21	36
		Ariel Hinojosa Subdivision	Already served under City of Mission (North Mission).									
		Barney Groves	5	5.000	5	0	0	5		1	1	12
		Basham # 1	Already served under City of Mission (North Mission).									
		Basham # 2	Already served under City of Mission (North Mission).									
		Basham # 4	Already served under City of Mission (North Mission).									
		Basham # 5	Already served under City of Mission (North Mission).									
		Basham # 6	Already served under City of Mission (North Mission).									
		Basham # 7	Already served under City of Mission (North Mission).									
		Basham # 8	Already served under City of Mission (North Mission).									
		Basham #11	Already served under City of Mission (North Mission).									
		Basham #12	59	4.538	59	0	0	59		13	13	28
		Basham #13	50	4.545	50	0	0	50		11	11	28
		Basham #15	108	4.500	108	0	0	108		24	24	41
		Basham #16	144	4.500	144	0	0	144		32	32	53

Continued

Table A.10 (continued)
Hidalgo County Project Summary

Project Status	Project Name / Sponsor	Colonia Name	# of Residents	Density	# Svd. by Public Water	# Not Svd. By Public Water	# Svd. By Central Waste-water	# Not Svd. By Central Waste-water	# Svd. By Project	# of Dwellings	# of Occupied Lots	Total # of Lots
Facilities Planning (continued)		Basham #18	77	4.529	77	0	0	77		17	17	29
		Basham Mobile Home Subd #19	131	4.517	131	0	0	131		29	29	33
		Bentsen Palm RV Park No 2	14	4.667	14	0	0	14		3	3	3
		Breyfogle Park Subd #1	27	4.500	27	0	0	27		6	6	10
		Canadiana Estates	59	4.538	59	0	0	59		13	13	42
		Carlos Leal Jr Subd #2	Already served under City of Mission (North Mission).									
		Carlos Subdivision	176	4.513	176	0	0	176		39	39	43
		Carol Subdivision	23	4.600	23	0	0	23		5	5	26
		Cavazos, Alex	Already served under City of Mission (North Mission).									
		Celso	108	4.500	108	0	0	108		24	24	24
		Chacon Estates	Already served under City of Mission (North Mission).									
		Country Corner Estates	77	4.529	77	0	0	77		17	17	19
		Country Estates West	Already served under City of Mission (North Mission).									
		Country Estates West Addn A	18	4.500	18	0	0	18		4	4	6
		Country Grove Estates	108	4.500	108	0	0	108		24	24	32
		Cuatro Vientos	113	4.520	113	0	0	113		25	25	40
		Del Norte	Already served under City of Mission (North Mission).									
		Diamond L #2	81	4.500	81	0	0	81		18	18	42
		Dina's Subdivision	32	4.571	32	0	0	32		7	7	11
		Enrique Bazan Subdivision	104	4.522	104	0	0	104		23	23	30
		Expressway Acres	99	4.500	99	0	0	99		22	19	19
		Ezequiel Acevedo	140	4.516	140	0	0	140		31	26	26

Continued

Table A.10 (continued)
Hidalgo County Project Summary

Project Status	Project Name / Sponsor	Colonia Name	# of Residents	Density	# Svd. by Public Water	# Not Svd. By Public Water	# Svd. By Central Waste-water	# Not Svd. By Central Waste-water	# Svd. By Project	# of Dwellings	# of Occupied Lots	Total # of Lots
Facilities Planning (continued)		Ezequiel Acevedo Jr Subd #2	86	4.526	86	0	0	86		19	19	19
		Four Sure All Right #1	95	4.524	95	0	0	95		21	21	56
		Francis	50	4.545	50	0	0	50		11	11	15
		G & R Subd	5	5.000	5	0	0	5		1	1	5
		Garza Estates	45	4.500	45	0	0	45		10	10	10
		Gomez Subdivision	14	4.667	14	0	0	14		3	3	3
		Good Valley Ranch	Already served under City of Mission (North Mission).									
		Goodwin Acres #1	23	4.600	23	0	0	23		5	5	5
		Goodwin Acres #2 (Amended)	59	4.538	59	0	0	59		13	13	20
		Goodwin Acres #3	104	4.522	104	0	0	104		23	23	24
		Goodwin Heights #1	252	4.500	252	0	0	252		56	56	88
		Goodwin West Subd #1	41	4.556	41	0	0	41		9	9	11
		Goodwin West Subd #2	72	4.500	72	0	0	72		16	16	49
		Goodwin West Subd #3	18	4.500	18	0	0	18		4	4	5
		Hilda	333	4.500	333	0	0	333		74	74	89
		Hilda #2	Already served under City of Mission (North Mission).									
		Hilda #3	68	4.533	68	0	0	68		15	15	22
		Hill-Top Subdivision	18	4.500	18	0	0	18		4	4	5
		Inspiration Point Subd	32	4.571	0	32	0	32		7	7	7
		J & O Subdivision	162	4.500	162	0	0	162		36	36	45
		Kountry Hill Estates	189	4.500	189	0	0	189		42	42	56
		Kristi Estates	Already served under City of Mission (North Mission).									
		La Camellia	54	4.500	54	0	0	54		12	12	34
		La Camellia A	99	4.500	99	0	0	99		22	22	49
		La Homa Acres	Already served under City of Mission (North Mission).									

Continued

Table A.10 (continued)
Hidalgo County Project Summary

Project Status	Project Name / Sponsor	Colonia Name	# of Residents	Density	# Svd. by Public Water	# Not Svd. By Public Water	# Svd. By Central Waste-water	# Not Svd. By Central Waste-water	# Svd. By Project	# of Dwellings	# of Occupied Lots	Total # of Lots
Facilities Planning (continued)		La Homa Acres #2	Already served under City of Mission (North Mission).									
		La Homa Acres #4	32	4.571	32	0	0	32		7	7	11
		La Homa Five Subdivision	27	4.500	27	0	0	27		6	6	31
		La Homa Grove Estates	Already served under City of Mission (North Mission).									
		La Homa Grove Estates No 2	50	4.545	50	0	0	50		11	11	13
		La Homa Road	153	4.500	153	0	0	153		34	34	52
		La Homa Road North (Amended)	Already served under City of Mission (North Mission).									
		La Palma	Already served under City of Mission (North Mission).									
		La Paloma Site	261	4.500	261	0	0	261		58	58	95
		La Suena	Already served under City of Mission (North Mission).									
		Lakeside	32	4.571	32	0	0	32		7	7	12
		Los Trevinos Subd #2	45	4.500	45	0	0	45		10	10	10
		Los Trevinos Subd #3	36	4.500	36	0	0	36		8	8	10
		Los Trevinos Subd #4	135	4.500	135	0	0	135		30	30	36
		Los Trevinos Subd #5	212	4.511	212	0	0	212		47	47	48
		Los Trevinos Subdivision	9	4.500	9	0	0	9		2	2	2
		Loya	32	4.571	32	0	0	32		7	7	12
		M & S Subdivision	Already served under City of Mission (North Mission).									
		Maier	23	4.600	23	0	0	23		5	5	13
		Marla Subdivision	54	4.500	54	0	0	54		12	12	38
		Mary K Acres	23	4.600	23	0	0	23		5	5	9
		Mesquite Ridge	54	4.500	54	0	0	54		12	12	20
		Mission West Estates	14	4.667	14	0	0	14		3	3	41
		Moorefield Acres	27	4.500	27	0	0	27		6	6	13
		Moorefield Groves Estates	5	5.000	5	0	0	5		1	1	5

Continued

Table A.10 (continued)
Hidalgo County Project Summary

Project Status	Project Name / Sponsor	Colonia Name	# of Residents	Density	# Svd. by Public Water	# Not Svd. By Public Water	# Svd. By Central Waste-water	# Not Svd. By Central Waste-water	# Svd. By Project	# of Dwellings	# of Occupied Lots	Total # of Lots
Facilities Planning (continued)		Munoz Estates	41	4.556	41	0	0	41		9	9	69
		Nick Garza Subdivi.	90	4.500	90	0	0	90		20	20	24
		North Country Estates #2	50	4.545	50	0	0	50		11	11	24
		Orleander Estates	9	4.500	9	0	0	9		2	2	17
		Palm Acres Estates	45	4.500	45	0	0	45		10	10	16
		Palm Creek	99	4.500	99	0	0	99		22	22	47
		Palm Drive North #1	Already served under City of Mission (North Mission).									
		Palma Alta	36	4.500	36	0	0	36		8	8	14
		Palmerina	36	4.500	36	0	0	36		8	8	14
		Palmview Paradise	95	4.524	95	0	0	95		21	21	36
		Park Lane	72	4.500	72	0	0	72		16	16	28
		Paseo De Palmas	32	4.571	32	0	0	32		7	7	20
		Que Pasa Subd	27	4.500	27	0	0	27		6	6	54
		R Ruiz	36	4.500	36	0	0	36		8	8	20
		Ramirez Estates	126	4.500	126	0	0	126		28	28	34
		Ramirez Subd	18	4.500	18	0	0	18		4	4	20
		Ramirez Subd No 2	68	4.533	68	0	0	68		15	15	20
		Ramirez Subd No 3	54	4.500	54	0	0	54		12	12	24
		Ramirez Subd No 4	54	4.500	54	0	0	54		12	12	20
		Randolph Barnett # 1	Already served under City of Mission (North Mission).									
		Randolph Barnett # 2	72	4.500	72	0	0	72		16	16	22
		Regal Estates										
		Resub Plat of Jimenez Subd	113	4.520	113	0	0	113		25	24	24
		Royal Palm Estates	5	5.000	5	0	0	5		1	1	50
		Schuerbach Acres	Already served under City of Mission (North Mission).									
		Sno-bird Estates	117	4.500	117	0	0	117		26	26	56
		Sno-Bird Estates #2	9	4.500	9	0	0	9		2	2	20
		Sotira Estates	72	4.500	72	0	0	72		16	16	55

Continued

Table A.10 (continued)
Hidalgo County Project Summary

Project Status	Project Name / Sponsor	Colonia Name	# of Residents	Density	# Svd. by Public Water	# Not Svd. By Public Water	# Svd. By Central Waste-water	# Not Svd. By Central Waste-water	# Svd. By Project	# of Dwellings	# of Occupied Lots	Total # of Lots
Facilities Planning (continued)		South Minnesota Road Subd 1	99	4.500	99	0	0	99		22	22	36
		South Minnesota Road Subd 2	63	4.500	63	0	0	63		14	14	24
		South Minnesota Road Subd 3	72	4.500	72	0	0	72		16	16	28
		Sun Valley	77	4.529	77	0	0	77		17	17	28
		Sunny Haven Estates	122	4.519	122	0	0	122		27	27	56
		Thompson	Already served under City of Mission (North Mission).									
		Tierra Linda	329	4.507	329	0	0	329		73	73	335
		Tommy Knocker	9	4.500	9	0	0	9		2	2	2
		Valle Grande Subdivision #2	63	4.500	63	0	0	63		14	14	34
		Valle Hermoso Estates	45	4.500	45	0	0	45		10	10	10
		Villa Capri	Already served under City of Mission (North Mission).									
		West Highway	32	4.571	32	0	0	32		7	7	19
		Westview Heights	5	5.000	5	0	0	5		1	1	14
	TOTAL - Palmview		**8,061**	**4.511**	**8,029**	**32**	**0**	**8,061**	**8,061**	**1,787**	**1,778**	**3,365**
	Penitas											
		Colonia Camargo	63	4.500	0	63	0	63		14	14	14
		Colonia Martinez	95	4.524	95	0	0	95		21	21	31
		Daniel Ozuna	189	4.500	189	0	0	189		42	42	52
		El Rio	149	4.515	149	0	0	149		33	33	57
		King Ranch #1	122	4.519	122	0	0	122		27	27	42
		King Ranch #2	180	4.500	180	0	0	180		40	40	70
		Penitas	653	4.503	653	0	0	653		145	145	145
		Penitas Nuevo	288	4.500	288	0	0	288		64	64	93
		Puerta Blanca										
		Ramona	45	4.500	0	45	0	45		10	10	16

Continued

Table A.10 (continued)
Hidalgo County Project Summary

Project Status	Project Name / Sponsor	Colonia Name	# of Residents	Density	# Svd. by Public Water	# Not Svd. By Public Water	# Svd. By Central Waste-water	# Not Svd. By Central Waste-water	# Svd. By Project	# of Dwellings	# of Occupied Lots	Total # of Lots
Facilities Planning (continued)		Reina	32	4.571	0	32	0	32		7	7	11
	TOTAL - Penitas		**1,816**	**4.506**	**1,676**	**140**	**0**	**1,816**	**1,816**	**403**	**403**	**531**
	San Juan											
		Aldamas Unit	275	4.508	275	0	0	275		61	61	85
		Arco Iris #2	284	4.508	284	0	0	284		63	63	78
		Arguello #1	63	4.500	63	0	0	63		14	14	24
		Arguello #2	86	4.526	86	0	0	86		19	19	29
		Azteca Acres	131	4.517	131	0	0	131		29	29	58
		Bar Unit #4	396	4.500	396	0	0	396		88	88	134
		Barrios #2	23	4.600	23	0	0	23		5	5	8
		Border Breeze	86	4.526	86	0	0	86		19	19	32
		Citriana Village	162	4.500	162	0	0	162		36	36	79
		Eldora Gardens	68	4.533	68	0	0	68		15	15	17
		Las Brisas	Already served under City of Pharr (Las Milpas).									
		Las Palmas Est	261	4.500	261	0	0	261		58	58	90
		Miller Re Subd Lot A	63	4.500	63	0	0	63		14	14	24
		Morningside Mobil Hm Park	54	4.500	54	0	0	54		12	12	18
		Morningside Terrace	324	4.500	324	0	0	324		72	72	120
		Morningsun	144	4.500	144	0	0	144		32	32	42
		Paradise Park	131	4.517	131	0	0	131		29	29	46
		Porciones Center Subd	5	5.000	5	0	0	5		1	1	1
		Primavera I	563	4.504	563	0	0	563		125	125	180
		Primavera II (Phase 2)	162	4.500	162	0	0	162		36	36	53
		R.S.W. Unit #1	225	4.500	225	0	0	225		50	37	37
		Red Barn	99	4.500	99	0	0	99		22	22	36
		Romo Subdivision	18	4.500	18	0	0	18		4	4	4
		San Juan East	162	4.500	162	0	0	162		36	36	36
		Sioux Terrace	401	4.506	401	0	0	401		89	89	150

Continued

Table A.10 (continued)
Hidalgo County Project Summary

Project Status	Project Name / Sponsor	Colonia Name	# of Residents	Density	# Svd. by Public Water	# Not Svd. By Public Water	# Svd. By Central Waste-water	# Not Svd. By Central Waste-water	# Svd. By Project	# of Dwellings	# of Occupied Lots	Total # of Lots
Facilities Planning (continued)		Sioux Terrace South	360	4.500	360	0	0	360		80	80	106
		Sun Valley Est #1	140	4.516	140	0	0	140		31	31	54
		Sundowners Retire	Already served under HC - Urban Regional WW Planning.									
		Tower Lodge	50	4.545	50	0	0	50		11	11	24
		Unnamed (Raul Longoria)	189	4.500	189	0	0	189		42	42	43
		Valley Star Acres	Already served under HC - Urban Regional WW Planning.									
	TOTAL - San Juan		**4,925**	**4.506**	**4,925**	**0**	**0**	**4,925**	**4,925**	**1,093**	**1,080**	**1,608**
	Hidalgo County - El Paraiso											
		Casa De Los Vecinos	293	4.508	293	0	0	293		65	65	96
		El Paraiso	167	4.514	167	0	0	167		37	37	50
		El Paraiso Surrounding	58	4.462	58	0	0	58		13	13	13
		Los Ebanos 1 & 2	774	4.500	774	0	0	774		172	172	172
	TOTAL - Hidalgo County - El Paraiso		**1,292**	**4.502**	**1,292**	**0**	**0**	**1,292**	**1,292**	**287**	**287**	**331**
EDAP TOTAL* HIDALGO COUNTY		**Number of Colonias: 583**	**80,782**	**4.612**	**78,526**	**2,256**	**1,296**	**79,486**	**80,782**	**17,516**	**17,707**	**26,912**
	*** Not including City of Mission (North Mission) project. There is no cost estimate available.**											
Colonias Not Served in EDAP												
		11 North/Victoria Rd-FM 493	36	4.500	0	36	0	36		8	8	8
		12 1/2 North/FM 88	140	4.516	0	140	0	140		31	31	31
		13 1/2 North/FM 493	23	4.600	0	23	0	23		5	5	5
		13 1/2 North/FM 493	18	4.500	0	18	0	18		4	4	10
		13 North/2 West	36	4.500	0	36	0	36		8	8	8
		15 1/2 North/FM 491	22	4.400	0	22	0	22		5	5	5
		17 1/2 North/6 West	63	4.500	0	63	0	63		14	14	14

Continued

Table A.10 (continued)
Hidalgo County Project Summary

Project Status	Project Name / Sponsor	Colonia Name	# of Residents	Density	# Svd. by Public Water	# Not Svd. By Public Water	# Svd. By Central Waste-water	# Not Svd. By Central Waste-water	# Svd. By Project	# of Dwellings	# of Occupied Lots	Total # of Lots
Colonias Not Served in EDAP (continued)		2812 Subdivision	15	5.000	0	15	0	15		3	3	3
		56	315	4.500	0	315	0	315		70	70	70
		7th Street Subdivision	15	5.000	0	15	0	15		3	3	3
		9 North/East FM 493	59	4.538	0	59	0	59		13	13	13
		Abram	734	4.503	734	0	0	734		163	163	163
		Acacia	54	4.500	54	0	0	54		12	12	20
		Acre Tract	72	4.500	72	0	0	72		16	16	27
		Adam Lee	9	4.500	0	9	0	9		2	2	30
		Ala Blanca #1 thru 4	140	4.516	140	0	0	140		31	31	52
		Ala Blanca Norte 1 thru 4	531	4.500	531	0	0	531		118	118	196
		Alamo Rose RV Resort	644	4.503	644	0	0	644		143	143	238
		Alamo RV Park	432	4.500	432	0	0	432		96	96	160
		Alberta/I Road	45	4.500	0	45	0	45		10	10	10
		Alsonia	734	4.503	0	734	0	734		163	163	272
		Altamira West	32	4.571	32	0	0	32		7	7	11
		Amigo Park I & II	337	4.493	337	0	0	337		75	75	60
		Amigo Park III	113	4.520	113	0	0	113		25	25	24
		Anaqua	36	4.500	36	0	0	36		8	8	13
		Anderson Rd/FM 493	23	4.600	0	23	0	23		5	5	5
		Avocado Park	104	4.522	104	0	0	104		23	23	38
		B & P Bridge (Toluca Ranch)	27	4.500	27	0	0	27		6	6	10
		B J Bonham	32	4.571	0	32	0	32		7	7	10
		Balli Subd.	189	4.500	189	0	0	189		42	42	52
		Barbosa-Lopez	426	5.071	426	0	0	426		84	84	187
		Basham #3	27	4.500	0	27	0	27		6	6	10
		Bensen (Unrecorded)	27	4.500	0	27	0	27		6	7	7
		Bernal	81	4.500	81	0	0	81		18	18	21
		Bertha	9	4.500	9	0	0	9		2	2	10

Continued

Table A.10 (continued)
Hidalgo County Project Summary

Project Status	Project Name / Sponsor	Colonia Name	# of Residents	Density	# Svd. by Public Water	# Not Svd. By Public Water	# Svd. By Central Waste-water	# Not Svd. By Central Waste-water	# Svd. By Project	# of Dwellings	# of Occupied Lots	Total # of Lots
Colonias Not Served in EDAP (continued)		Betos Acres	54	4.500	54	0	0	54		12	12	21
		Big 5 Rd # 5 (Unrecorded)	90	4.500	0	90	0	90		20	20	20
		Boderland Subd.	77	2.852	77	0	0	77		27	17	142
		Boyd #1	14	4.667	14	0	0	14		3	3	47
		Bryan (Unrecorded)	18	4.500	0	18	0	18		4	4	4
		Bustamante (Olympia)	36	4.500	36	0	0	36		8	8	56
		Campo Alto	908	4.586	908	0	0	908		198	198	206
		Capetillo	131	4.517	131	0	0	131		29	29	48
		Capisallo (Rio Grande)	545	4.504	545	0	0	545		121	121	158
		Carlos Acres	135	4.500	135	0	0	135		30	30	78
		Casa Bonita	50	4.545	50	0	0	50		11	11	30
		Catherine	68	4.533	68	0	0	68		15	15	21
		Cerrito	54	4.500	54	0	0	54		12	12	30
		Chapa #1	36	4.500	36	0	0	36		8	8	40
		Chapa #2	268	4.323	268	0	0	268		62	62	76
		Chapa, Josefina L	68	4.533	68	0	0	68		15	15	10
		Cinco Hermanas	59	4.214	59	0	0	59		14	14	22
		Citrus Retreat	50	4.545	50	0	0	50		11	11	18
		CJRS Unit A & B	23	4.600	23	0	0	23		5	5	28
		Col Garza	90	4.500	90	0	0	90		20	20	20
		Collin	59	4.538	59	0	0	59		13	13	26
		Colonia #1	37	4.111	37	0	0	37		9	9	10
		Colonia #2	25	5.000	25	0	0	25		5	5	13
		Colonia Allende	18	4.500	18	0	0	18		4	4	9
		Colonia Claude Lookingbill	176	4.513	176	0	0	176		39	39	59
		Colonia Delmiro Jackson	50	4.545	0	50	0	50		11	11	20

Continued

Table A.10 (continued)
Hidalgo County Project Summary

Project Status	Project Name / Sponsor	Colonia Name	# of Residents	Density	# Svd. by Public Water	# Not Svd. By Public Water	# Svd. By Central Waste-water	# Not Svd. By Central Waste-water	# Svd. By Project	# of Dwellings	# of Occupied Lots	Total # of Lots
Colonias Not Served in EDAP (continued)		Colonia Evans	185	4.512	185	0	0	185		41	41	47
		Colonia George	5	5.000	0	5	0	5		1	1	8
		Colonia Las Palmas	414	4.500	0	414	0	414		92	92	153
		Colonia Lucero del Norte	243	4.500	243	0	0	243		54	54	90
		Colonia Noreste	666	4.500	666	0	0	666		148	148	385
		Colonia Saenz	59	4.538	0	59	0	59		13	13	18
		Colonia Victoriana	27	4.500	27	0	0	27		6	6	10
		Conway	23	4.600	0	23	0	23		5	5	24
		Cottonwood	72	4.500	72	0	0	72		16	16	26
		Country Acres	27	4.500	27	0	0	27		6	6	10
		Country Air Est	27	4.500	27	0	0	27		6	6	10
		Country Air Est #4	131	4.517	131	0	0	131		29	29	48
		Country Living Est	113	4.520	113	0	0	113		25	25	42
		Country View	90	4.500	90	0	0	90		20	20	33
		Dellinger	23	4.600	23	0	0	23		5	5	8
		Delta Courts	193	5.079	193	0	0	193		38	38	16
		Delta II	54	4.500	54	0	0	54		12	8	10
		Delta West	234	4.500	234	0	0	234		52	52	74
		Delta/Rodger Subdivision	23	4.600	0	23	0	23		5	5	5
		Diana	86	4.526	86	0	0	86		19	19	77
		Dimas #2 & 3	131	4.517	131	0	0	131		29	29	48
		Ebony Hollow	113	4.520	113	0	0	113		25	25	41
		El Charro	567	4.500	567	0	0	567		126	126	153
		El Nopal	113	4.520	0	113	0	113		25	25	34
		Eldora Rd/FM 1426	104	4.522	0	104	0	104		23	23	23
		Eldora/Tower	44	5.500	0	44	0	44		8	8	8
		Enchanted Valley Ranch	1,274	4.502	1,274	0	0	1,274		283	283	384
		Encino	261	4.500	261	0	0	261		58	58	97

Continued

Table A.10 (continued)
Hidalgo County Project Summary

Project Status	Project Name / Sponsor	Colonia Name	# of Residents	Density	# Svd. by Public Water	# Not Svd. By Public Water	# Svd. By Central Waste-water	# Not Svd. By Central Waste-water	# Svd. By Project	# of Dwellings	# of Occupied Lots	Total # of Lots
Colonias Not Served in EDAP (continued)		Encino Heights	252	4.500	252	0	0	252		56	56	97
		Engleman Estates	162	4.500	162	0	0	162		36	36	86
		Estates, 281	171	4.500	171	0	0	171		38	38	44
		Evangeline Gardens	86	4.526	86	0	0	86		19	19	42
		Expressway 83/Ricardo Rd	99	4.500	0	99	0	99		22	22	22
		Expresswway Hghts	724	4.090	724	0	0	724		177	177	189
		Flea Market Subdivision	50	4.545	0	50	0	50		11	11	15
		Flora	228	4.560	228	0	0	228		50	50	58
		FM 1426/Minn Rd	32	4.571	0	32	0	32		7	7	7
		FM 1925/Floral Rd	45	4.500	0	45	0	45		10	10	10
		Foster	63	4.500	63	0	0	63		14	14	24
		Gate City	54	4.500	54	0	0	54		12	12	12
		Gernentz	423	4.500	423	0	0	423		94	94	141
		Glasscock North	36	4.500	36	0	0	36		8	8	14
		Gray East & West	32	4.571	32	0	0	32		7	7	12
		Green Valley Dev	36	4.500	36	0	0	36		8	8	19
		Guerra-Ellis Subdivision	32	4.571	32	0	0	32		7	7	14
		H & B Subd.	77	4.529	77	0	0	77		17	17	20
		H. M. E.	158	4.514	158	0	0	158		35	35	51
		Hacienda Del Porvenir	23	4.600	0	23	0	23		5	5	9
		Harmel	27	4.500	27	0	0	27		6	6	13
		High Point	162	4.500	162	0	0	162		36	36	40
		Highland Heights	122	4.519	0	122	0	122		27	27	37
		Highview	261	4.500	261	0	0	261		58	58	96
		Hillcrest Terrace	140	4.516	140	0	0	140		31	31	58
		Hoehn Drive (Unrecorded)	14	4.667	0	14	0	14		3	3	3

Continued

Table A.10 (continued)
Hidalgo County Project Summary

Project Status	Project Name / Sponsor	Colonia Name	# of Residents	Density	# Svd. by Public Water	# Not Svd. By Public Water	# Svd. By Central Waste-water	# Not Svd. By Central Waste-water	# Svd. By Project	# of Dwellings	# of Occupied Lots	Total # of Lots
Colonias Not Served in EDAP (continued)		I Rd/Minn Rd	99	4.500	0	99	0	99		22	22	22
		I Rd/Owasso-Kennedy	45	4.500	0	45	0	45		10	10	10
		Ignacio Perez	32	4.571	32	0	0	32		7	7	12
		Imperial	50	4.545	0	50	0	50		11	11	19
		Ingle-Doolittle	36	4.500	0	36	0	36		8	8	8
		Inspiration Rd #1 thru 3	5	5.000	5	0	0	5		1	1	7
		Jackson (Unrecorded)	68	4.533	0	68	0	68		15	15	80
		James-Allen Subdivision	54	4.500	54	0	0	54		12	12	64
		Jesus Maria	131	4.517	131	0	0	131		29	29	48
		JR # 1 & 2	45	4.500	45	0	0	45		10	10	22
		Jussup Rollo (Monte Alto)	153	4.500	0	153	0	153		34	34	57
		Kaufold #1	41	4.556	41	0	0	41		9	9	14
		Kenyan Hgts.	113	4.520	113	0	0	113		25	25	27
		L C Olivarez	40	4.444	0	40	0	40		9	9	14
		La Blanca Est	185	4.512	185	0	0	185		41	41	98
		La Casa Real #1	212	4.511	212	0	0	212		47	47	79
		La Estancia	30	2.000	30	0	0	30		15	15	35
		La Frontera	203	4.511	203	0	0	203		45	45	75
		La Hacienda	50	4.545	50	0	0	50		11	11	18
		La Hacienda Est	302	4.507	302	0	0	302		67	67	97
		La Homa Groves Subd 1, 2 & 3	27	4.500	27	0	0	27		6	6	19
		La Loma Alta Subd	340	4.857	340	0	0	340		70	70	102
		La Palma #1	815	5.292	815	0	0	815		154	154	96
		La Pampa	54	4.500	54	0	0	54		12	12	16
		La Reyna	135	4.500	135	0	0	135		30	30	123
		LaComa Heights	167	4.514	167	0	0	167		37	37	61
		Laguna Park	239	4.509	239	0	0	239		53	53	74

Continued

Table A.10 (continued)
Hidalgo County Project Summary

Project Status	Project Name / Sponsor	Colonia Name	# of Residents	Density	# Svd. by Public Water	# Not Svd. By Public Water	# Svd. By Central Waste-water	# Not Svd. By Central Waste-water	# Svd. By Project	# of Dwellings	# of Occupied Lots	Total # of Lots
Colonias Not Served in EDAP (continued)		Lake View Subd.	68	4.533	68	0	0	68		15	15	78
		Lane & Redgate	18	4.500	18	0	0	18		4	4	42
		Lanfranco	54	4.500	54	0	0	54		12	12	12
		Lantana Meadows	63	4.500	63	0	0	63		14	14	24
		Las Brisas del Sur	59	4.538	59	0	0	59		13	13	45
		Le Leona	131	4.517	131	0	0	131		29	29	49
		Loma Chica	45	4.500	45	0	0	45		10	10	17
		Loma Linda Hghts.	140	4.516	140	0	0	140		31	31	88
		Long & Green Park	59	4.538	59	0	0	59		13	13	21
		Lopezville	495	4.500	495	0	0	495		110	110	141
		Los Ebanos (Unrecorded)	23	4.600	0	23	0	23		5	5	50
		Los Leones	117	4.500	117	0	0	117		26	26	43
		Lotts	54	4.500	0	54	0	54		12	11	13
		Lyons	63	4.500	63	0	0	63		14	14	24
		M & R	36	4.500	0	36	0	36		8	8	13
		Magnolia Unit 1	72	4.500	0	72	0	72		16	16	16
		Martin	68	4.533	68	0	0	68		15	15	27
		Mary Ann	95	4.524	95	0	0	95		21	21	32
		Mata #2	153	4.500	153	0	0	153		34	34	57
		McColl	221	4.510	221	0	0	221		49	49	49
		McKee	113	4.520	113	0	0	113		25	25	37
		Meadow Creek Country Club	873	4.500	873	0	0	873		194	194	323
		Mel Gray	162	4.500	0	162	0	162		36	36	36
		Mesquite	86	4.526	86	0	0	86		19	19	33
		Mid-Valley Estates	246	5.467	246	0	0	246		45	37	62
		Mid-Way Village	96	2.743	96	0	0	96		35	35	75
		Minn Road	99	4.500	0	99	0	99		22	22	22
		Monte Cristo Heights	45	4.500	45	0	0	45		10	10	16
		Monte Cristo Hills	234	4.500	234	0	0	234		52	52	86

Continued

Table A.10 (continued)
Hidalgo County Project Summary

Project Status	Project Name / Sponsor	Colonia Name	# of Residents	Density	# Svd. by Public Water	# Not Svd. By Public Water	# Svd. By Central Waste-water	# Not Svd. By Central Waste-water	# Svd. By Project	# of Dwellings	# of Occupied Lots	Total # of Lots
Colonias Not Served in EDAP (continued)		Nelle Est	23	4.600	23	0	0	23		5	5	9
		New Palms	50	4.545	50	0	0	50		11	11	18
		North Alamo Est.	117	4.500	117	0	0	117		26	26	32
		North Capisallo	63	4.500	63	0	0	63		14	14	14
		North Country Estates #1	158	4.514	158	0	0	158		35	35	44
		North Santa Cruz Subd	32	4.571	0	32	0	32		7	7	7
		Northern Acres	32	4.571	32	0	0	32		7	7	8
		O & J	203	4.511	203	0	0	203		45	45	99
		Old Rebel Field	45	4.500	0	45	0	45		10	10	17
		Olivarez #1	54	4.500	54	0	0	54		12	12	24
		Olivarez #2	58	4.462	0	58	0	58		13	13	18
		Olivarez #3	45	4.500	0	45	0	45		10	10	17
		Olivarez (12 North)	152	4.222	152	0	0	152		36	36	34
		Olivarez Tr 304	89	4.450	0	89	0	89		20	17	20
		Orchard Homes Add #2	131	4.517	131	0	0	131		29	29	39
		Oriente	50	4.545	50	0	0	50		11	11	12
		Ortega, Thomas Subd	27	4.500	27	0	0	27		6	6	11
		Owasso Rd/Tower Rd	63	4.500	0	63	0	63		14	14	14
		Owasso/Bus 281	77	4.529	0	77	0	77		17	17	17
		Palma #1	207	4.500	207	0	0	207		46	46	81
		Palmhurst Manor #1	63	4.500	63	0	0	63		14	14	14
		Palmview	72	4.500	72	0	0	72		16	16	26
		Parajitos	36	4.500	36	0	0	36		8	8	10
		Piquito Oro	113	4.520	113	0	0	113		25	25	177
		Pleasant Valley Ranch	203	4.511	203	0	0	203		45	45	361
		Praille	27	4.500	27	0	0	27		6	6	14
		Ramerez A & E #1	68	4.533	68	0	0	68		15	15	25
		Ramerez A&E #2	90	4.500	90	0	0	90		20	20	33

Continued

Table A.10 (continued)
Hidalgo County Project Summary

Project Status	Project Name / Sponsor	Colonia Name	# of Residents	Density	# Svd. by Public Water	# Not Svd. By Public Water	# Svd. By Central Waste-water	# Not Svd. By Central Waste-water	# Svd. By Project	# of Dwellings	# of Occupied Lots	Total # of Lots
Colonias Not Served in EDAP (continued)		Ranchette Est	63	4.500	63	0	0	63		14	14	46
		Rancho Escondido	288	4.500	288	0	0	288		64	64	85
		Rankin	72	4.500	72	0	0	72		16	16	35
		Raquet Club	77	4.529	77	0	0	77		17	17	29
		Re Subd Lot 14, Blk 145	270	4.500	270	0	0	270		60	60	100
		Rebecca	23	4.600	0	23	0	23		5	5	8
		Regency Acres	324	4.500	324	0	0	324		72	72	120
		Relampago	117	4.500	117	0	0	117		26	26	26
		Restful Valley Ranch	891	4.500	891	0	0	891		198	198	423
		Rice	108	4.500	108	0	0	108		24	24	40
		Riverside Est.	338	4.507	338	0	0	338		75	75	100
		Robinette	36	4.500	36	0	0	36		8	8	13
		Rodgers Rd Subdivision	36	4.500	0	36	0	36		8	8	8
		Rodriguez	180	4.500	180	0	0	180		40	40	70
		Roosevelt/FM 1423	63	4.500	0	63	0	63		14	14	14
		Rosa Linda Subd.	86	4.526	86	0	0	86		19	19	35
		Rosedale Heights	135	4.821	135	0	0	135		28	28	42
		Royalty House #2 & 3	329	4.507	329	0	0	329		73	73	122
		Runn	63	4.500	63	0	0	63		14	14	14
		San Carlos Acres	86	4.526	86	0	0	86		19	19	22
		San Carlos Farms	41	4.556	41	0	0	41		9	9	12
		San Juan	126	4.500	126	0	0	126		28	28	46
		San Juan South Est.	126	4.500	126	0	0	126		28	28	46
		Santa Amalia	45	4.500	45	0	0	45		10	10	28
		Schunior	104	4.522	104	0	0	104		23	23	29
		Seminary Est.	81	4.500	81	0	0	81		18	18	30
		Seminary Village	126	4.500	126	0	0	126		28	28	47
		Seville Park #1	86	4.526	86	0	0	86		19	19	22

Continued

Table A.10 (continued)
Hidalgo County Project Summary

Project Status	Project Name / Sponsor	Colonia Name	# of Residents	Density	# Svd. by Public Water	# Not Svd. By Public Water	# Svd. By Central Waste-water	# Not Svd. By Central Waste-water	# Svd. By Project	# of Dwellings	# of Occupied Lots	Total # of Lots
Colonias Not Served in EDAP (continued)		SH 88/12 North/4 1/2 West	45	4.500	0	45	0	45		10	10	10
		SH 88/14 North/6 West	81	4.500	0	81	0	81		18	18	18
		SH 88/15 North/4 West	36	4.500	0	36	0	36		8	8	8
		Shary (Unrecorded)	23	4.600	0	23	0	23		5	5	5
		Shull	144	4.500	144	0	0	144		32	32	44
		Siesta Village #1,2,3 & 4	1,845	4.500	1,845	0	0	1,845		410	410	410
		Siez Tract	36	4.500	36	0	0	36		8	8	14
		Silverado	95	4.524	95	0	0	95		21	21	45
		Sing	338	4.507	0	338	0	338		75	75	75
		Sioux/Morningside	22	4.400	0	22	0	22		5	5	5
		South Palm Garden Est #1&2	81	4.500	81	0	0	81		18	18	30
		South Seminary	99	4.500	99	0	0	99		22	22	36
		Southern Breeze	212	4.511	212	0	0	212		47	47	68
		Spring Gardens	63	4.500	63	0	0	63		14	14	29
		Stephensons	90	4.500	0	90	0	90		20	20	25
		Sugar (Unrecorded)	68	4.533	0	68	0	68		15	15	15
		Sun Country Est.	300	4.054	300	0	0	300		74	74	103
		Sun Valley Est.	135	4.500	135	0	0	135		30	30	37
		Sun Valley Estates	153	4.500	153	0	0	153		34	34	57
		Sunset Est	23	4.600	23	0	0	23		5	5	8
		Sunshine Valley Est	540	4.500	540	0	0	540		120	120	200
		Sylvia	27	4.500	27	0	0	27		6	6	10
		Taylor Rd (N3)	36	4.500	0	36	0	36		8	8	8
		Taylor Rd (N4)	45	4.500	0	45	0	45		10	10	16
		Texas/Morningside-Doolittle	40	4.444	0	40	0	40		9	9	9

Continued

Table A.10 (continued)
Hidalgo County Project Summary

Project Status	Project Name / Sponsor	Colonia Name	# of Residents	Density	# Svd. by Public Water	# Not Svd. By Public Water	# Svd. By Central Waste-water	# Not Svd. By Central Waste-water	# Svd. By Project	# of Dwellings	# of Occupied Lots	Total # of Lots
Colonias Not Served in EDAP (continued)		Tiejerina Estates	108	4.500	0	108	0	108		24	24	24
		Tierra Dorado #1 & 2	1,364	4.502	1,364	0	0	1,364		303	303	747
		Tierra Maria	198	4.500	198	0	0	198		44	44	54
		Timberhill Villa	419	4.505	419	0	0	419		93	93	153
		Tiny Acres	149	4.515	149	0	0	149		33	33	38
		Todd #1, 2, & 3	230	4.510	230	0	0	230		51	51	59
		Tom Gill Road	68	4.533	0	68	0	68		15	15	22
		Tower Acres #1	131	4.517	131	0	0	131		29	29	48
		Tropicana	77	4.529	77	0	0	77		17	17	26
		Trosper (Unrecorded)	45	4.500	0	45	0	45		10	10	10
		Uvalde	32	4.571	0	32	0	32		7	7	12
		Val Bar Est	81	4.500	81	0	0	81		18	18	40
		Valle Alto #1 & 2	572	4.504	572	0	0	572		127	127	164
		Valle de Palmas #1	144	4.500	144	0	0	144		32	32	49
		Valley Rancheros	144	4.500	144	0	0	144		32	32	54
		Valley View Est	99	4.500	99	0	0	99		22	22	38
		Vertriss	72	4.500	72	0	0	72		16	16	26
		Victoria Bend	59	4.538	59	0	0	59		13	13	22
		Villa Verde	936	4.500	936	0	0	936		208	208	220
		Ware	23	4.600	0	23	0	23		5	5	16
		Water Fall Road	108	4.500	108	0	0	108		24	24	164
		Weather Heights (Ph 1)	63	4.500	0	63	0	63		14	14	24
		Wes-Mer	570	4.914	570	0	0	570		116	116	134
		Westgate	32	4.571	0	32	0	32		7	7	16
		Whalen & 83	166	4.486	0	166	0	166		37	37	37
		Wildwood Forest	72	4.500	72	0	0	72		16	16	27
		Wisconsin/Dillon Rds	23	4.600	0	23	0	23		5	5	5
		Wisconsin/I Road	49	4.455	0	49	0	49		11	11	11
		Woods	482	4.505	482	0	0	482		107	107	40
		Yoakum Hall	86	4.526	86	0	0	86		19	19	32

Continued

Table A.10 (continued)
Hidalgo County Project Summary

Project Status	Project Name / Sponsor	Colonia Name	# of Residents	Density	# Svd. by Public Water	# Not Svd. By Public Water	# Svd. By Central Waste-water	# Not Svd. By Central Waste-water	# Svd. By Project	# of Dwellings	# of Occupied Lots	Total # of Lots
Colonias Not Served in EDAP (continued)		Zacatal	87	4.579	87	0	0	87		19	19	19
		Zambrows Subdivision	58	4.462	0	58	0	58		13	13	13
		Zuley Manor	27	4.500	27	0	0	27		6	6	18
NOT SERVED BY EDAP - TOTAL HIDALGO COUNTY		**Number of Colonias: 285**	**43,228**	**4.513**	**37,302**	**5,926**	**0**	**43,228**	**0**	**9,578**	**9,553**	**15,305**

Table A.11
Hudspeth County Project Summary

Project Status	Project Name / Sponsor	Colonia Name	# of Residents	Density	# Svd. by Public Water	# Not Svd. by Public Water	# Svd. by Central Waste-water	# Not Svd. by Central Waste-water	# Svd. by Project	# of Dwellings	# of Occupied Lots	Total # of Lots
Plans and Specifications												
	Hudspeth WCID #1											
		Sierra Blanca	887	4.503	887	0	0	887		197	197	350
	TOTAL - Hudspeth WCID #1		**887**	**4.503**	**887**	**0**	**0**	**887**	**887**	**197**	**197**	**350**
EDAP TOTAL HUDSPETH COUNTY		**Number of Colonias: 1**	**887**	**4.503**	**887**	**0**	**0**	**887**	**887**	**197**	**197**	**350**
Colonias Not Served in EDAP												
		Acala	95	4.524	0	95	0	95	0	21	21	21
		Villa Alegre	36	4.500	36	0	0	36	0	8	8	8
NOT SERVED BY EDAP - TOTAL HUDSPETH COUNTY		**Number of Colonias: 2**	**131**	**4.517**	**36**	**95**	**0**	**131**	**1,774**	**29**	**29**	**29**

Table A.12
Jeff Davis County Project Summary

Project Status	Project Name / Sponsor	Colonia Name	# of Residents	Density	# Svd. by Public Water	# Not Svd. by Public Water	# Svd. by Central Waste-water	# Not Svd. by Central Waste-water	# Svd. by Project	# of Dwellings	# of Occupied Lots	Total # of Lots
Colonias Not Served in EDAP												
		Valentine	200	2.353	200	0	0	200		85	85	400
NOT SERVED BY EDAP - TOTAL JEFF DAVIS COUNTY		**Number of Colonias: 1**	**200**	**2.353**	**200**	**0**	**0**	**200**	**0**	**85**	**85**	**400**

Table A.13
Jim Hogg County Project Summary

Project Status	Project Name / Sponsor	Colonia Name	# of Residents	Density	# Svd. by Public Water	# Not Svd. by Public Water	# Svd. by Central Waste-water	# Not Svd. by Central Waste-water	# Svd. by Project	# of Dwellings	# of Occupied Lots	Total # of Lots
Colonias Not Served in EDAP												
		Guerra	30	2.000	0	30	0	30		15	17	17
		Highway 16 South	23	3.833	0	23	0	23		6	6	30
		Los Lomitas	77	3.850	58	19	0	77		20	20	40
NOT SERVED BY EDAP - TOTAL JIM HOGG COUNTY		**Number of Colonias: 3**	**130**	**3.171**	**58**	**72**	**0**	**130**	**0**	**41**	**43**	**87**

Table A.14
Jim Wells County Project Summary

Project Status	Project Name / Sponsor	Colonia Name	# of Residents	Density	# Svd. by Public Water	# Not Svd. by Public Water	# Svd. by Central Waste-water	# Not Svd. by Central Waste-water	# Svd. by Project	# of Dwellings	# of Occupied Lots	Total # of Lots
Colonias Not Served in EDAP												
		665 Site	200	4.444	0	200	0	200		45	45	45
		Alice Acres	126	4.500	0	126	0	126		28	28	49
		Amargosa	257	4.509	0	257	0	257		57	57	84
		Collins Townsite	90	4.500	0	90	0	90		20	20	20
		Coyote Acres	176	4.513	0	176	0	176		39	39	45
		English Acres	149	4.515	0	149	0	149		33	33	52
		Hilltop Acres	185	4.512	0	185	0	185		41	41	60
		K-Bar	324	4.500	0	324	0	324		72	72	99
		Kiesling	180	4.000	0	180	0	180		45	45	45
		Loma Linda Subdivision	167	4.514	0	167	0	167		37	37	37
		Orange Grove Villa	100	3.125	0	100	0	100		32	32	40
		Rancho Alegre	2,100	3.500	2,100	0	1,400	700		600	600	1,320
		Salinas Subdivision	50	4.545	0	50	0	50		11	11	11
		San Petronila	180	4.500	0	180	0	180		40	40	100
		Sandia	900	4.000	0	900	0	900		225	225	225
		Tecolote Subdivision	392	4.506	0	392	0	392		87	87	115
NOT SERVED BY EDAP - TOTAL JIM WELLS COUNTY		**Number of Colonias: 16**	**5,576**	**3.949**	**2,100**	**3,476**	**1,400**	**4,176**	**0**	**1,412**	**1,412**	**2,347**

Table A.15
Kinney County Project Summary

Project Status	Project Name / Sponsor	Colonia Name	# of Residents	Density	# Svd. by Public Water	# Not Svd. by Public Water	# Svd. by Central Waste-water	# Not Svd. by Central Waste-water	# Svd. by Project	# of Dwellings	# of Occupied Lots	Total # of Lots
Facilities Planning												
	Spofford											
		Spofford	81	2.793	81	0	0	81		29	29	29
	TOTAL - Spofford		**81**	**2.793**	**81**	**0**	**0**	**81**	**81**	**29**	**29**	**29**
EDAP TOTAL KINNEY COUNTY		**Number of Colonias: 1**	**81**	**2.793**	**81**	**0**	**0**	**81**	**81**	**29**	**29**	**29**
Colonias Not Served in EDAP												
		Brackettville	250	4.167	250	0	0	250		60	60	160
NOT SERVED BY EDAP - TOTAL KINNEY COUNTY		**Number of Colonias: 1**	**250**	**4.167**	**250**	**0**	**0**	**250**	**0**	**60**	**60**	**160**

Table A.16
La Salle County Project Summary

Project Status	Project Name / Sponsor	Colonia Name	# of Residents	Density	# Svd. by Public Water	# Not Svd. by Public Water	# Svd. by Central Waste-water	# Not Svd. by Central Waste-water	# Svd. by Project	# of Dwellings	# of Occupied Lots	Total # of Lots
Facilities Planning												
	La Salle County											
		Encinal	960	4.000	960	0	0	960		240	220	700
	TOTAL - La Salle County		**960**	**4.000**	**960**	**0**	**0**	**960**	**960**	**240**	**220**	**700**
Colonis Not Served in EDAP												
		Artesia Wells	60	5.000	0	60	0	60		12	12	12
		Fowlerton	200	4.000	0	200	0	200		50	60	2,000
		Gardendale	92	4.000	0	92	0	92		23	23	500
		Los Angeles	76	4.471	0	76	0	76		17	9	400
		Millet	49	3.500	0	49	0	49		14	14	102
		Zamora Acres	28	4.000	12	16	0	28		7	6	10
NOT SERVED BY EDAP - TOTAL LA SALLE COUNTY		**Number of Colonias: 6**	**505**	**4.106**	**12**	**493**	**0**	**505**	**0**	**123**	**124**	**3,024**

Table A.17
Maverick County Project Summary

Project Status	Project Name / Sponsor	Colonia Name	# of Residents	Density	# Svd. by Public Water	# Not Svd. by Public Water	# Svd. by Central Waste-water	# Not Svd. by Central Waste-water	# Svd. by Project	# of Dwellings	# of Occupied Lots	Total # of Lots
Under Construction												
	City of Eagle Pass											
		Eagle Heights	135	4.500	135	0	0	135		30	30	294
		Green Acres	68	4.533	68	0	0	68		15	15	56
		Heritage Farms	95	4.524	95	0	0	95		21	21	29
		Lago Vista	104	4.522	104	0	0	104		23	23	39
		Las Brisas	846	4.500	846	0	0	846		188	188	261
		Las Quintas Fronterizas	3,195	4.500	3,195	0	0	3,195		710	710	716
		Loma Bonita	2,664	4.500	2,664	0	0	2,664		592	592	949
		Morales	360	4.500	360	0	0	360		80	80	80
	TOTAL - City of Eagle Pass		**7,467**	**4.501**	**7,467**	**0**	**0**	**7,467**	**7,467**	**1,659**	**1,659**	**2,424**
Facilities Planning												
	Quernado											
		Normandy	81	4.500	0	81	0	81		18	18	41
		Quemado	450	4.500	0	450	0	450		100	100	358
	TOTAL - Quernado		**531**	**4.500**	**0**	**531**	**0**	**531**	**531**	**118**	**118**	**399**
EDAP TOTAL MAVERICK COUNTY		**Number of Colonias: 10**	**7,998**	**4.501**	**7,467**	**531**	**0**	**7,998**	**7,998**	**1,777**	**1,777**	**2,823**
Colonias Not Served in EDAP												
		Airport Addition	149	4.515	149	0	0	149		33	33	133
		Big River Park	27	4.500	27	0	0	27		6	6	15
		Cedar Ridge I&II	18	4.500	18	0	0	18		4	4	42
		Chula Vista 1-5	1,778	4.501	1,778	0	0	1,778		395	395	597
		Chula Vista School Blk	135	4.500	135	0	0	135		30	30	57
		Deer Run 1-5	360	4.500	360	0	0	360		80	80	632
		El Indio Townsite	342	4.500	342	0	0	342		76	76	300

Continued

Table A.17 (continued)
Maverick County Project Summary

Project Status	Project Name / Sponsor	Colonia Name	# of Residents	Density	# Svd. by Public Water	# Not Svd. by Public Water	# Svd. by Central Waste-water	# Not Svd. by Central Waste-water	# Svd. by Project	# of Dwellings	# of Occupied Lots	Total # of Lots
Colonias Not Served in EDAP (continued)		Elm Creek 1 & 2	180	4.500	180	0	0	180		40	40	91
		Florentine Ramos	18	4.500	18	0	0	18		4	4	6
		Hector Rodriguez	9	4.500	9	0	0	9		2	2	11
		Hopedale	50	4.545	50	0	0	50		11	11	76
		Jose Roberto Rodriguez	72	4.500	72	0	0	72		16	16	31
		Kickapoo Indian Village	324	4.500	324	0	0	324		72	72	132
		La Herradura	180	4.500	180	0	0	180		40	40	203
		Las Carretas	248	4.509	248	0	0	248		55	55	104
		Las Hacienditas	288	4.500	288	0	0	288		64	64	104
		Loma Linda	18	4.500	18	0	0	18		4	4	387
		Loma Linda Ranchetts	360	4.500	360	0	0	360		80	80	371
		Los Guajillos	36	4.500	36	0	0	36		8	8	78
		Los Jardines Verdes	81	4.500	81	0	0	81		18	18	33
		Morales Circle	14	4.667	14	0	0	14		3	3	17
		Nellis Lands	104	4.522	104	0	0	104		23	23	23
		Paisano Heights	18	4.500	18	0	0	18		4	4	75
		Pueblo Nuevo	54	4.500	54	0	0	54		12	12	456
		Radar Base	122	4.519	122	0	0	122		27	27	27
		Riverside Acres	248	4.509	248	0	0	248		55	55	218
		Rockaway Country Site	36	4.500	36	0	0	36		8	8	25
		Rosita Gardens	63	4.500	63	0	0	63		14	14	155
		Rosita Valley	396	4.500	396	0	0	396		88	88	125
		Sauz Creek	99	4.500	99	0	0	99		22	22	25
		South Elm Creek	18	4.500	18	0	0	18		4	4	59
		Victoriano Hernandez	54	4.500	54	0	0	54		12	12	68
		Wilson & Bargo	45	4.500	45	0	0	45		10	10	17
		Zamora Lands	27	4.500	27	0	0	27		6	6	10
NOT SERVED BY EDAP - TOTAL MAVERICK COUNTY		**Number of Colonias: 34**	**5,971**	**4.503**	**5,971**	**0**	**0**	**5,971**	**0**	**1,326**	**1,326**	**4,703**

Table A.18
Newton County Project Summary

Project Status	Project Name / Sponsor	Colonia Name	# of Residents	Density	# Svd. by Public Water	# Not Svd. by Public Water	# Svd. by Central Waste-water	# Not Svd. by Central Waste-water	# Svd. by Project	# of Dwellings	# of Occupied Lots	Total # of Lots
Colonias Not Served in EDAP												
		Bleakwood	630	3.500	0	630	0	630		180	180	180
		Bon Weir	600	3.488	600	0	0	600		172	172	172
		Deweyville	3,750	3.498	3,750	0	0	3,750		1,072	1,072	1,072
		Old Salem	240	3.478	0	240	0	240		69	69	69
		Toledo Village	2,230	3.501	2,230	0	0	2,230		637	637	637
		Trout Creek	510	3.469	0	510	0	510		147	147	147
NOT SERVED BY EDAP - TOTAL NEWTON COUNTY		**Number of Colonias: 6**	**7,960**	**3.496**	**6,580**	**1,380**	**0**	**7,960**	**0**	**2,277**	**2,277**	**2,277**

Table A.19
Pecos County Project Summary

Project Status	Project Name / Sponsor	Colonia Name	# of Residents	Density	# Svd. by Public Water	# Not Svd. by Public Water	# Svd. by Central Waste-water	# Not Svd. by Central Waste-water	# Svd. by Project	# of Dwellings	# of Occupied Lots	Total # of Lots
Colonias Not Served in EDAP												
		Alamo Ranchetts	322	2.556	0	322	0	322		126	126	150
		Bodieville	205	3.796	205	0	0	205		54	54	64
		Coyanosa	200	3.333	186	14	0	200		60	65	260
		Imperial	363	2.174	363	0	0	363		167	175	1,714
		Little Mexico	360	4.000	320	40	0	360		90	90	120
NOT SERVED BY EDAP - TOTAL PECOS COUNTY		**Number of Colonias: 5**	**1,450**	**2.918**	**1,074**	**376**	**0**	**1,450**	**0**	**497**	**510**	**2,308**

Table A.20
Presidio County Project Summary

Project Status	Project Name / Sponsor	Colonia Name	# of Residents	Density	# Svd. by Public Water	# Not Svd. by Public Water	# Svd. by Central Waste-water	# Not Svd. by Central Waste-water	# Svd. by Project	# of Dwellings	# of Occupied Lots	Total # of Lots
Facilities Planning												
	Ruidosa											
		Ruidosa	59	5.900	0	59	0	59		10	10	25
	TOTAL - Ruidosa		**59**	**5.900**	**0**	**59**	**0**	**59**	**59**	**10**	**10**	**25**
EDAP TOTAL PRESISIO COUNTY		**Number of Colonias: 1**	**59**	**5.900**	**0**	**59**	**0**	**59**	**59**	**10**	**10**	**25**
Colonias Not Served in EDAP												
		Candelaria	100	4.348	0	100	0	100		23	23	26
		Las Pampas	60	4.000	0	60	0	60		15	15	17
		Loma Pelona	24	3.000	0	24	0	24		8	8	8
		Pueblo Nuevo	320	4.000	320	0	0	320		80	80	580
		Redford	163	3.881	0	163	0	163		42	42	150
		Shafter	30	2.143	0	30	0	30		14	14	16
NOT SERVED BY EDAP - TOTAL PRESIDIO COUNTY		**Number of Colonias: 6**	**697**	**3.830**	**320**	**377**	**0**	**697**	**0**	**182**	**182**	**797**

Table A.21
Red River County Project Summary

Project Status	Project Name / Sponsor	Colonia Name	# of Residents	Density	# Svd. by Public Water	# Not Svd. by Public Water	# Svd. by Central Waste-water	# Not Svd. by Central Waste-water	# Svd. by Project	# of Dwellings	# of Occupied Lots	Total # of Lots
Colonias Not Served in EDAP												
		Annona	63	2.333	63	0	37	26		27	27	27
		Avery	42	3.500	42	0	0	42		12	12	12
		Bagwell	225	3.000	225	0	0	225		75	75	75
		Clarksville FM 2825	52	4.000	52	0	0	52		13	13	13
		Clarksville Hwy 1159	40	4.000	40	0	0	40		10	10	10
		Clarksville Hwy 37	40	4.000	40	0	0	40		10	10	10
		Clarksville Hwy 82 West	12	4.000	12	0	0	12		3	3	3
		Clarksville Hwy 910	48	4.000	48	0	0	48		12	12	12
		Detroit	128	4.000	128	0	0	128		32	32	32
		Lydia	36	3.000	0	36	0	36		12	12	12
		Town North & Casa Linda	200	6.250	200	0	0	200		32	32	50
NOT SERVED BY EDAP - TOTAL RED RIVER COUNTY		**Number of Colonias: 11**	**886**	**3.723**	**850**	**36**	**37**	**849**	**0**	**238**	**238**	**256**

Table A.22
Reeves County Project Summary

Project Status	Project Name / Sponsor	Colonia Name	# of Residents	Density	# Svd. by Public Water	# Not Svd. by Public Water	# Svd. by Central Waste-water	# Not Svd. by Central Waste-water	# Svd. by Project	# of Dwellings	# of Occupied Lots	Total # of Lots
Colonias Not Served in EDAP												
		Lindsay Division	400	3.390	400	0	0	400		118	118	159
		Toyah	140	2.500	140	0	140	0		56	56	56
NOT SERVED BY EDAP - TOTAL REEVES COUNTY		**Number of Colonias: 2**	**540**	**3.103**	**540**	**0**	**140**	**400**	**0**	**174**	**174**	**215**

Table A.23
Sabine County Project Summary

Project Status	Project Name / Sponsor	Colonia Name	# of Residents	Density	# Svd. by Public Water	# Not Svd. by Public Water	# Svd. by Central Waste-water	# Not Svd. by Central Waste-water	# Svd. by Project	# of Dwellings	# of Occupied Lots	Total # of Lots
Colonias Not Served in EDAP												
		Bronson	327	2.019	327	0	0	327		162	162	162
		Brookeland	242	2.017	242	0	0	242		120	120	120
		Delta Heights	49	2.333	49	0	0	49		21	21	21
		Highway 96	63	2.864	63	0	0	63		22	22	22
		Toldeo Bend	6,294	2.020	6,294	0	0	6,294		3,116	3,116	3,116
NOT SERVED BY EDAP - TOTAL SABINE COUNTY		**Number of Colonias: 5**	**6,975**	**2.027**	**6,975**	**0**	**0**	**6,975**	**0**	**3,441**	**3,441**	**3,441**

Table A.24
San Patricio County Project Summary

Project Status	Project Name / Sponsor	Colonia Name	# of Residents	Density	# Svd. by Public Water	# Not Svd. by Public Water	# Svd. by Central Waste-water	# Not Svd. by Central Waste-water	# Svd. by Project	# of Dwellings	# of Occupied Lots	Total # of Lots
Facilities Planning												
	Ingleside											
		Ingleside	Information not available on the Texas Water Devlopment Board's Needs Database as of December, 1995.									
	TOTAL - Ingelside		**Information not available on the Texas Water Devlopment Board's Needs Database as of December, 1995.**									
	San Patricio County											
		Bethel Estates	450	3.982	450	0	0	450		113	113	113
		Doyle Addition	260	4.000	260	0	0	260		65	65	65
		Edroy	500	1.370	500	0	500	0		365	365	365
		Edroy-Sinton Rural Area	1,528	3.497	0	1,528	0	1,528		437	437	437
		Gregory	2,458	3.501	2,458	0	2,458	0		702	702	702
		Lake City	465	1.449	0	465	0	465		321	321	321
		Lakeside	292	1.587	0	292	0	292		184	184	184
		N Lake Shore G/Hidden Acres	506	2.530	0	506	0	506		200	200	200
		Saint Paul	345	2.654	345	0	0	345		130	135	135
		South Taft	765	4.005	765	0	0	765		191	191	191
		Southeast Odem	100	3.333	100	0	0	100		30	30	30
		West Aransas Pass	1,761	3.000	0	1,761	0	1,761		587	587	587
	TOTAL - San Patricio County		**9,430**	**2.836**	**4,878**	**4,552**	**2,958**	**6,472**	**9,430***	**3,325**	**3,330**	**3,330**
		* Based on the assumption that the people who need water are not the same as those who need wastewater infrastructure. There will be some overlap.										
EDAP TOTAL SAN PATRICIO COUNTY**		**Number of Colonias: 13**	**9,430**	**2.836**	**4,878**	**4,552**	**2,958**	**6,472**	**9,430**	**3,325**	**3,330**	**3,330**
		** Not including Ingleside project. Informaiton on Ingleside colonia not available on the Texas Water Development Board's needs database.										
Colonias Not Served in EDAP												
		Buena Vistas/Bueno Vista	120	4.000	120	0	120	0		30	30	30
		Dodd	151	3.512	151	0	151	0		43	43	43

Continued

Table A.24 (continued)
San Patricio County Project Summary

Project Status	Project Name / Sponsor	Colonia Name	# of Residents	Density	# Svd. by Public Water	# Not Svd. by Public Water	# Svd. by Central Waste-water	# Not Svd. by Central Waste-water	# Svd. by Project	# of Dwellings	# of Occupied Lots	Total # of Lots
Colonias Not Served in EDAP (continued)		J. G. Gonzales	249	4.016	249	0	249	0		62	62	62
		La Colonia	310	3.974	310	0	0	310		78	78	78
		Rancho Chico	330	3.976	330	0	330	0		83	83	83
		Tradewinds	246	3.000	0	246	0	246		82	82	82
NOT SERVED BY EDAP - TOTAL SAN PATRICIO COUNTY		**Number of Colonias: 6**	**1,406**	**3.720**	**1,160**	**246**	**850**	**556**	**0**	**378**	**378**	**378**

Table A.25
Starr County Project Summary

Project Status	Project Name / Sponsor	Colonia Name	# of Residents	Density	# Svd. by Public Water	# Not Svd. by Public Water	# Svd. by Central Waste-water	# Not Svd. by Central Waste-water	# Svd. by Project	# of Dwellings	# of Occupied Lots	Total # of Lots
Plans and Specifications												
	Starr Co. WCID #2 (Las Lomas)											
		Las Lomas	2,066	3.791	2,066	0	0	2,066		545	545	600
	TOTAL - Starr Co. WCID #2 (Las Lomas)		**2,066**	**3.791**	**2,066**	**0**	**0**	**2,066**	**2,066**	**545**	**545**	**600**
Facilities Planning												
	Roma											
		Amando Pena # 1	185	4.512	185	0	0	185		41	41	41
		Amando Pena # 2	45	4.500	45	0	0	45		10	10	10
		Cantu	50	4.545	50	0	0	50		11	11	11
		De La Cruz (Roma)	135	4.655	135	0	0	135		29	29	29
		El Bosque # 1	131	4.517	131	0	0	131		29	29	29
		El Bosque # 2	162	4.500	162	0	0	162		36	36	36
		El Bosque # 3	212	4.511	212	0	0	212		47	47	47
		El Bosque # 4	158	4.514	158	0	0	158		35	35	35
		Escobares	135	4.500	135	0	0	135		30	30	30
		Escobares # 1	491	4.505	491	0	0	491		109	109	109
		Fronton	225	4.500	225	0	0	225		50	50	50
		Hackberry	153	4.500	153	0	0	153		34	34	34
		Hillside Terrace	0	0.000	0	0	0	0		0	0	62
		J. F. Villarreal	86	4.526	86	0	0	86		19	19	40
		Las Flores	243	4.500	243	0	0	243		54	54	54
		Loma Vista	329	4.507	329	0	0	329		73	73	73
		Loma Vista # 1	167	4.514	167	0	0	167		37	37	37
		Mesquite # 1	342	4.500	342	0	0	342		76	76	76
		Mesquite # 2	405	4.500	405	0	0	405		90	90	90
		Mesquite # 3	221	4.510	221	0	0	221		49	49	49
		Mesquite # 4	135	4.500	135	0	0	135		30	30	30
		Mirasoles	113	4.520	113	0	0	113		25	25	55

Continued

Table A.25 (continued)
Starr County Project Summary

Project Status	Project Name / Sponsor	Colonia Name	# of Residents	Density	# Svd. by Public Water	# Not Svd. by Public Water	# Svd. by Central Waste-water	# Not Svd. by Central Waste-water	# Svd. by Project	# of Dwellings	# of Occupied Lots	Total # of Lots
Facilities Planning (continued)		Munoz	90	4.500	90	0	0	90		20	20	20
		Munoz-Garcia	351	4.500	351	0	0	351		78	78	78
		Rivera Subdivision	144	4.500	144	0	0	144		32	48	56
		Robinson	243	4.500	243	0	0	243		54	54	54
		Santa Catalina	203	4.511	203	0	0	203		45	45	45
		Victoria	158	4.514	158	0	0	158		35	35	150
	TOTAL - Roma		**5,312**	**4.509**	**5,312**	**0**	**0**	**5,312**	**5,312**	**1,178**	**1,194**	**1,430**
EDAP TOTAL STARR COUNTY		**Number of Colonias: 29**	**7,378**	**4.282**	**7,378**	**0**	**0**	**7,378**	**7,378**	**1,723**	**1,739**	**2,030**
Colonias Not Served in EDAP												
		A. T. Martinez	18	4.500	18	0	0	18		4	4	4
		Alto Bonito Heights	554	4.504	554	0	0	554		123	123	123
		Alto Bonito Hghts. # 5	180	4.500	180	0	0	180		40	40	40
		Arroyo Ranch	405	5.063	405	0	0	405		80	80	80
		Benito Saenz	356	4.506	356	0	0	356		79	87	113
		Berrera	135	4.500	135	0	0	135		30	30	60
		Buena Vista	36	4.500	36	0	0	36		8	8	41
		Campo Verde	50	4.545	50	0	0	50		11	17	41
		Chapa	2,493	4.500	2,493	0	0	2,493		554	554	554
		Cortez	432	4.500	432	0	0	432		96	96	96
		Cuellar Estates	162	4.500	162	0	0	162		36	36	36
		De La Garza Subdivision	284	4.508	284	0	0	284		63	63	63
		De Los Santos Subd	54	4.500	54	0	0	54		12	15	26
		Doyno West Side # 2	96	6.000	96	0	25	71		16	17	122
		East Alto Bonito	1,625	4.501	1,625	0	0	1,625		361	361	361
		El Castillo Subdivision	158	4.514	158	0	0	158		35	37	49
		El Cenizo	315	4.500	315	0	0	315		70	70	70

Continued

Table A.25 (continued)
Starr County Project Summary

Project Status	Project Name / Sponsor	Colonia Name	# of Residents	Density	# Svd. by Public Water	# Not Svd. by Public Water	# Svd. by Central Waste-water	# Not Svd. by Central Waste-water	# Svd. by Project	# of Dwellings	# of Occupied Lots	Total # of Lots
Colonias Not Served in EDAP (continued)		El Chaparral # 1	419	4.505	419	0	0	419		93	93	93
		El Chaparral # 2	230	4.510	230	0	0	230		51	51	51
		El Mesquite	135	4.500	135	0	0	135		30	30	30
		El Quiote 1 & 2	200	3.922	200	0	0	200		51	53	55
		El Rancho Vela	410	4.505	410	0	0	410		91	91	91
		El Refugio Subdivision	50	4.545	50	0	0	50		11	11	19
		El Socio Subdivision	81	4.500	81	0	0	81		18	20	40
		Elodia's	63	4.500	63	0	0	63		14	14	20
		Eugenio Saenz Subd.	162	4.500	162	0	0	162		36	38	60
		Evergreen	99	4.500	99	0	0	99		22	22	65
		Falconaire Subd 1 & 2	108	4.500	108	0	0	108		24	31	308
		Flor Del Rio Subd	60	4.000	60	0	0	60		15	20	26
		Fourth Site Subdivision	45	4.500	45	0	0	45		10	12	15
		Garceno	270	4.500	270	0	0	270		60	60	60
		Garcia	257	4.509	257	0	0	257		57	57	57
		Garza-Salinas Subd	540	4.500	540	0	0	540		120	129	160
		Gloria	27	4.500	0	27	0	27		6	6	42
		Guadalupe Guerra Subd	27	4.500	27	0	0	27		6	8	56
		Hill Top	140	4.516	140	0	0	140		31	31	31
		La Escondida	243	4.500	243	0	0	243		54	54	54
		La Esperanza	585	4.500	585	0	0	585		130	130	130
		La Lomita	162	4.500	162	0	0	162		36	36	36
		La Puerta Subdivision	369	4.500	369	0	0	369		82	85	100
		Las Palmas	90	4.500	90	0	0	90		20	20	20
		Leal Subdivision	95	4.524	95	0	0	95		21	25	50
		Live Oak Est.	414	4.500	414	0	0	414		92	92	92
		Loma Linda	50	4.545	50	0	0	50		11	11	61
		Loma Linda Subd	36	4.500	36	0	0	36		8	8	14
		Loma Linda Subd Tr 2	36	4.500	36	0	0	36		8	8	14
		Longoria	75	5.000	75	0	0	75		15	18	28

Continued

Table A.25 (continued)
Starr County Project Summary

Project Status	Project Name / Sponsor	Colonia Name	# of Residents	Density	# Svd. by Public Water	# Not Svd. by Public Water	# Svd. by Central Waste-water	# Not Svd. by Central Waste-water	# Svd. by Project	# of Dwellings	# of Occupied Lots	Total # of Lots
Colonias Not Served in EDAP (continued)		Lopez-Chapa Unrecorded	104	4.522	104	0	0	104		23	26	28
		Los Alvarez	135	4.500	135	0	0	135		30	30	30
		Los Berreras	135	4.500	135	0	0	135		30	30	30
		Los Ebanos	536	4.504	536	0	0	536		119	119	119
		Los Morenos	158	4.514	158	0	0	158		35	35	35
		Manuel Gonzales	279	4.500	279	0	0	279		62	62	62
		Margarita Subdivision	68	4.533	68	0	0	68		15	17	35
		Meme # 1	279	4.500	279	0	0	279		62	70	75
		Miguel Barrera	90	4.500	90	0	0	90		20	20	53
		Mike's	1,571	4.501	1,571	0	0	1,571		349	349	349
		Montalvo Hills Subd	41	4.556	41	0	0	41		9	10	10
		Munoz	45	4.500	45	0	0	45		10	11	15
		North Escobares Ranchettes	81	4.500	0	81	0	81		18	20	320
		North Santa Cruz	905	4.502	905	0	0	905		201	201	201
		Olivare	266	4.508	266	0	0	266		59	59	59
		Olivia Guterrez	144	4.500	144	0	0	144		32	32	32
		Olmito	252	4.500	252	0	0	252		56	56	87
		Olmito # 2	972	4.500	972	0	0	972		216	216	216
		Pablo Blanco Subd.	200	4.444	200	0	0	200		45	51	91
		Pedro Compos Add.	77	4.529	77	0	0	77		17	17	19
		Pena Brothers Subds	185	5.000	185	0	0	185		37	49	60
		Quesada	36	4.500	36	0	0	36		8	8	22
		Rancho Viejo # 1	135	4.500	135	0	0	135		30	30	30
		Rancho Viejo # 2	113	4.520	113	0	0	113		25	25	25
		Raucon Drive Inn	104	4.522	104	0	0	104		23	23	24
		Regino Ramirez	81	4.500	81	0	0	81		18	18	51
		Reyna Subdivision	68	4.533	68	0	0	68		15	17	22
		Rivereno Subdivision	70	5.000	70	0	0	70		14	15	29
		Roma Cr # 1,2 & 3	216	4.500	216	0	0	216		48	48	134

Continued

Table A.25 (continued)
Starr County Project Summary

Project Status	Project Name / Sponsor	Colonia Name	# of Residents	Density	# Svd. by Public Water	# Not Svd. by Public Water	# Svd. by Central Waste-water	# Not Svd. by Central Waste-water	# Svd. by Project	# of Dwellings	# of Occupied Lots	Total # of Lots
Colonias Not Served in EDAP (continued)		Salineno	270	4.500	270	0	0	270		60	60	60
		Salmon	54	4.500	54	0	0	54		12	12	12
		San Isidro	1,098	4.500	0	1,098	0	1,098		244	244	244
		San Juan	135	4.500	135	0	0	135		30	38	60
		San Juan Subdivision	45	4.500	45	0	0	45		10	10	60
		Santa Cruz # 2	482	4.505	482	0	0	482		107	107	107
		Santa Cruz Ind. Park	1,197	4.500	1,197	0	0	1,197		266	266	266
		Santa Cruz Ind. Park # 4	261	4.500	261	0	0	261		58	58	58
		Santa Rosa	135	4.500	135	0	0	135		30	30	41
		Santel Subdivision	40	4.000	40	0	0	40		10	10	12
		Tamez	135	4.500	135	0	0	135		30	30	41
		Tierra Dorada	81	4.500	81	0	0	81		18	18	18
		Tierra Linda	365	4.506	365	0	0	365		81	81	81
		Trevinos	347	4.506	347	0	0	347		77	77	77
		Trevinos # 1	171	4.500	171	0	0	171		38	38	38
		Triple R	99	4.500	99	0	0	99		22	22	22
		Triple R Subdivision	23	4.600	23	0	0	23		5	5	50
		Valle Hermosa Subd	40	4.000	0	40	0	40		10	10	10
		Valle Vista # 1	401	4.506	401	0	0	401		89	89	89
		Valle Vista # 2	491	4.505	491	0	0	491		109	109	109
		Villareales	225	4.500	225	0	0	225		50	50	50
		Y. Saenz	140	4.516	140	0	0	140		31	31	54
		Zarate	54	4.500	54	0	0	54		12	15	23
NOT SERVED BY EDAP - TOTAL STARR COUNTY		**Number of Colonias: 99**	**26,466**	**4.512**	**25,220**	**1,246**	**25**	**26,441**	**0**	**5,866**	**5,976**	**7,602**

Table A.26
Terrell County Project Summary

Project Status	Project Name / Sponsor	Colonia Name	# of Residents	Density	# Svd. by Public Water	# Not Svd. by Public Water	# Svd. by Central Waste-water	# Not Svd. by Central Waste-water	# Svd. by Project	# of Dwellings	# of Occupied Lots	Total # of Lots
Facilities Planning												
	Terrell County WCID (Sanderson)											
		Sanderson	1,000	1.613	1,000	0	0	1,000	0	620	620	795
	TOTAL - Terrell County WCID (Sanderson)		**1,000**	**1.613**	**1,000**	**0**	**0**	**1,000**	**1,000**	**620**	**620**	**795**
EDAP TOTAL TERRELL COUNTY		**Number of Colonias: 1**	**1,000**	**1.613**	**1,000**	**0**	**0**	**1,000**	**1,000**	**620**	**620**	**795**

Table A.27
Uvalde County Project Summary

Project Status	Project Name / Sponsor	Colonia Name	# of Residents	Density	# Svd. by Public Water	# Not Svd. by Public Water	# Svd. by Central Waste-water	# Not Svd. by Central Waste-water	# Svd. by Project	# of Dwellings	# of Occupied Lots	Total # of Lots
Facilities Planning												
	Uvalde County											
		Brice Lane	78	3.900	78	0	0	78		20	20	24
		Fort Clark Road	31	2.818	31	0	0	31		11	10	50
		Knippa	600	2.390	600	0	0	600		251	251	300
		North Uvalde	397	3.609	397	0	0	397		110	30	60
		South Grove St	32	2.909	32	0	0	32		11	10	20
		Utopia	410	4.020	410	0	0	410		102		
		Uvalde Estates	500	1.779	500	0	0	500		281	175	423
		Vanessa Street	115	3.286	115	0	39	76		35	20	30
	TOTAL - Uvalde County		**2,163**	**2.635**	**2,163**	**0**	**39**	**2,124**	**2,124**	**821**	**516**	**907**
EDAP TOTAL UVALDE COUNTY		**Number of Colonias: 8**	**2,163**	**2.635**	**2,163**	**0**	**39**	**2,124**	**2,124**	**821**	**516**	**907**
Colonias Not Served in EDAP												
		Gonzales	110	4.583	110	0	110	0		24	24	24
NOT SERVED BY EDAP - TOTAL UVALDE COUNTY		**Number of Colonias: 1**	**110**	**4.583**	**110**	**0**	**110**	**0**	**0**	**24**	**24**	**24**

Table A.28
Val Verde County Project Summary

Project Status	Project Name / Sponsor	Colonia Name	# of Residents	Density	# Svd. by Public Water	# Not Svd. by Public Water	# Svd. by Central Waste-water	# Not Svd. by Central Waste-water	# Svd. by Project	# of Dwellings	# of Occupied Lots	Total # of Lots
Under Construction												
	City of Del Rio (Cienegas Terrace)											
		Cienegas Terrace	1,000	5.000	1,000	0	0	1,000		200	200	777
	TOTAL - City of Del Rio (Cienegas Terrace)		**1,000**	**5.000**	**1,000**	**0**	**0**	**1,000**	**1,000**	**200**	**200**	**777**
Plans and Specifications												
	City of Del Rio (Val Verde Park Estates)											
		Val Verde Park	450	4.500	450	0	0	450		100	100	1,236
	TOTAL - City of Del Rio (Val Verde Park Estates)		**450**	**4.500**	**450**	**0**	**0**	**450**	**450**	**100**	**100**	**1,236**
EDAP TOTAL VAL VERDE COUNTY		**Number of Colonias: 2**	**1,450**	**4.833**	**1,450**	**0**	**0**	**1,450**	**1,450**	**300**	**300**	**2,013**
Colonias Not Served in EDAP												
		Amistad Acres	200	2.020	0	200	0	200		99	100	100
		Box Canyon Estates	176	2.023	0	176	0	176		87	88	267
		Comstock	450	3.237	450	0	0	450		139	156	156
		Lake View Estates	400	3.030	0	400	0	400		132	135	246
		Langtry	51	3.000	51	0	0	51		17	32	32
		Los Campos	315	4.500	0	315	0	315		70	70	155
		Owens	90	4.500	90	0	0	90		20	20	38
		Payment	25	5.000	0	25	0	25		5	5	30
		Rio Bravo	310	4.493	310	0	0	310		69	69	69
NOT SERVED BY EDAP - TOTAL VAL VERDE COUNTY		**Number of Colonias: 9**	**2,017**	**3.161**	**901**	**1,116**	**0**	**2,017**	**0**	**638**	**675**	**1,093**

Table A.29
Webb County Project Summary

Project Status	Project Name / Sponsor	Colonia Name	# of Residents	Density	# Svd. by Public Water	# Not Svd. by Public Water	# Svd. by Central Waste-water	# Not Svd. by Central Waste-water	# Svd. by Project	# of Dwellings	# of Occupied Lots	Total # of Lots
Plans and Specifications												
	Webb County (Larga Vista)											
		Larga Vista	525	5.000	0	525	0	525		105	90	129
	TOTAL - Webb County (Larga Vista)		**525**	**5.000**	**0**	**525**	**0**	**525**	**525**	**105**	**90**	**129**
Facilities Planning												
	Webb County (Mines Road & S.H. 359)											
		Antonio Santos	76	3.800	0	76	0	76		20	20	20
		D-5 Acres	140	5.000	0	140	0	140		28	28	42
		La Coma	75	5.000	0	75	0	75		15	15	15
		Laredo Ranchettes	65	5.000	0	65	0	65		13	13	30
		Larga Vista										
		Los Altos I	500	5.000	0	500	0	500		100	75	95
		Los Corralitos	135	4.500	0	135	0	135		30	30	30
		Old Milwaukee	265	5.000	0	265	0	265		53	39	39
		Pueblo Nuevo	410	5.000	0	410	0	410		82	70	304
		Ranchitos 359 East	175	5.000	0	175	0	175		35	20	56
		Ranchitos IV Los Minerales	120	5.000	120	0	0	120		24	24	24
		Ranchos Penitas West	914	3.627	0	914	0	914		252	252	252
		San Carlos & San Enrique	600	5.000	0	600	0	600		120	100	160
		Tanquecitos S Acres I & II	295	5.000	0	295	0	295		59	40	106
	TOTAL - Webb County (Mines Road & S.H. 359)		**3,770**	**4.537**	**120**	**3,650**	**0**	**3,770**	**3,650***	**831**	**726**	**1,173**
	* EDAP notes indicate water service only. Residents in need of wastewater are not included in number of people served.											

Continued

Table A.29 (continued)
Webb County Project Summary

Project Status	Project Name / Sponsor	Colonia Name	# of Residents	Density	# Svd. by Public Water	# Not Svd. by Public Water	# Svd. by Central Waste-water	# Not Svd. by Central Waste-water	# Svd. by Project	# of Dwellings	# of Occupied Lots	Total # of Lots
Facilities Planning (continued)	Webb County (Southwest Webb)											
		El Cenizo	2,875	5.000	2,875	0	1,062	1,813		575	575	920
		La Presa	230	5.000	0	230	0	230		46	40	191
		One River Place	80	5.000	0	80	0	80		16	16	36
		Rio Bravo I, II, III, Annex	4,855	5.000	4,855	0	0	4,855		971	971	1,344
	TOTAL - Webb County (Southwest Webb)		**8,040**	**5.000**	**7,730**	**310**	**1,062**	**6,978**	**7,288****	**1,608**	**1,602**	**2,491**
	** Based on the assumption that people needing water are not the same as people needing wastewater.											
EDAP TOTAL* WEBB COUNTY**		**Number of Colonias: 18**	**4,295**	**4.589**	**120**	**4,175**	**0**	**4,295**	**4,175**	**936**	**816**	**1,302**
	*** Not including Webb County (Southwest Webb) project. There is no cost estimate available.											
Colonias Not Served in EDAP												
		Aguilares	75	5.000	0	75	0	75		15	15	32
		Bruni	645	5.000	645	0	0	645		129	105	105
		Colorado Acres	235	5.000	0	235	0	235		47	47	436
		East Gate Acres	30	5.000	0	30	0	30		6	6	70
		Four Points	30	5.000	0	30	0	30		6	6	34
		Hillside Acres I & II	60	5.000	0	60	0	60		12	12	150
		Las Lomas I & II	225	5.000	0	225	0	225		45	45	448
		Las Pilas I & II	75	5.000	0	75	0	75		15	15	186
		Los Arcos	95	5.000	0	95	0	95		19	19	129
		Los Botines	175	5.000	0	175	0	175		35	35	90
		Los Centenarios	85	5.000	0	85	0	85		17	17	111
		Los Fresnos	85	5.000	0	85	0	85		17	17	111
		Los Huisaches I & II	40	5.000	0	40	0	40		8	8	27
		Los Mesquites	40	5.000	0	40	0	40		8	8	56
		Los Napalias	70	5.000	0	70	0	70		14	14	94

Continued

Table A.29 (continued)
Webb County Project Summary

Project Status	Project Name / Sponsor	Colonia Name	# of Residents	Density	# Svd. by Public Water	# Not Svd. by Public Water	# Svd. by Central Waste-water	# Not Svd. by Central Waste-water	# Svd. by Project	# of Dwellings	# of Occupied Lots	Total # of Lots
Colonias Not Served in EDAP (continued)		Los Veteranos Ranchitos	25	5.000	0	25	0	25		5	5	24
		Los Veteranos Subdivision	69	3.833	0	69	0	69		18	18	18
		Mirando City	785	5.000	785	0	0	785		157	143	143
		Oilton & Rodriguez Addition	650	5.000	650	0	0	650		130	130	167
		Pueblo East	60	5.000	0	60	0	60		12	12	76
		Ranchitos Los Veteranos	324	4.500	0	324	0	324		72	72	72
		Regency Village	15	5.000	0	15	0	15		3	3	20
		Sunset Acres	45	5.000	0	45	0	45		9	9	20
		Valle Verde	65	5.000	0	65	0	65		13	13	85
		Village East	15	5.000	0	15	0	15		3	3	31
NOT SERVED BY EDAP - TOTAL WEBB COUNTY		**Number of Colonias: 25**	**4,018**	**4.930**	**2,080**	**1,938**	**0**	**4,018**	**15,638**	**815**	**777**	**2,735**

Table A.30
Willacy County Project Summary

Project Status	Project Name / Sponsor	Colonia Name	# of Residents	Density	# Svd. by Public Water	# Not Svd. by Public Water	# Svd. by Central Waste-water	# Not Svd. by Central Waste-water	# Svd. by Project	# of Dwellings	# of Occupied Lots	Total # of Lots
Complete												
	Sebastian WSC (Sebastian)											
		Sebastian	1,904	4.051	1,904	0	0	1,904		470	470	680
	TOTAL - Sebastian WSC (Sebastian)		**1,904**	**4.051**	**1,904**	**0**	**0**	**1,904**	**1,904**	**470**	**470**	**680**
Plans and Specifications												
	North Alamo WSC (La Sara)											
		La Sara	824	4.000	824	0	0	824		206	206	298
	TOTAL - North Alamo WSC (La Sara)		**824**	**4.000**	**824**	**0**	**0**	**824**	**824**	**206**	**206**	**298**
EDAP TOTAL WILLACY COUNTY		**Number of Colonias: 2**	**2,728**	**4.036**	**2,728**	**0**	**0**	**2,728**	**2,728**	**676**	**676**	**978**
Colonias Not Served in EDAP												
		Chapa Addition	90	4.500	90	0	0	90		20	20	46
		Colonia de Angeles	45	4.500	0	45	0	45		10	10	10
		Lyford North	130	3.824	130	0	0	130		34	34	49
		Lyford South	310	3.690	310	0	0	310		84	84	122
		Romoville	108	5.684	108	0	0	108		19	19	27
		Zapata Ranch	131	3.743	131	0	0	131		35	35	50
NOT SERVED BY EDAP - TOTAL WILLACY COUNTY		**Number of Colonias: 6**	**814**	**4.030**	**769**	**45**	**0**	**814**	**0**	**202**	**202**	**304**

Table A.31
Zapata County Project Summary

Project Status	Project Name / Sponsor	Colonia Name	# of Residents	Density	# Svd. by Public Water	# Not Svd. by Public Water	# Svd. by Central Waste-water	# Not Svd. by Central Waste-water	# Svd. by Project	# of Dwellings	# of Occupied Lots	Total # of Lots
Facilities Planning												
	Zapata County											
		Falcon Mesa	251	3.803	251	0	0	251		66	66	66
		Medina	1,991	3.800	1,991	0	0	1,991		524	524	1,048
	TOTAL - Zapata County		**2,242**	**3.800**	**2,242**	**0**	**0**	**2,242**	**2,242**	**590**	**590**	**1,114**
EDAP TOTAL ZAPATA COUNTY		**Number of Colonias: 2**	**2,242**	**3.800**	**2,242**	**0**	**0**	**2,242**	**2,242**	**590**	**590**	**1,114**
Colonias Not Served in EDAP												
		Cuellar	42	3.818	42	0	0	42		11	11	30
		Dolorosa	45	3.462	0	45	0	45		13	13	13
		Falcon	245	3.025	245	0	0	245		81	85	103
		Lopeno	280	4.118	280	0	0	280		68	75	92
		Siesta Shores	880	2.000	880	0	0	880		440	440	2,700
NOT SERVED BY EDAP - TOTAL ZAPATA COUNTY		**Number of Colonias: 5**	**1,492**	**2.434**	**1,447**	**45**	**0**	**1,492**	**0**	**613**	**624**	**2,938**

Table A.32
Zavala County Project Summary

Project Status	Project Name / Sponsor	Colonia Name	# of Residents	Density	# Svd. by Public Water	# Not Svd. by Public Water	# Svd. by Central Waste-water	# Not Svd. by Central Waste-water	# Svd. by Project	# of Dwellings	# of Occupied Lots	Total # of Lots
Facilities Planning												
	Zavala County (Batesville)											
		Batesville	1,600	4.482	1,600	0	0	1,600		357	357	357
	TOTAL - Zavala County (Batesville)		**1,600**	**4.482**	**1,600**	**0**	**0**	**1,600**	**1,600**	**357**	**357**	**357**
	Zavala County (La Pryor)											
		La Pryor	2,804	3.299	2,804	0	0	2,804		850	850	850
	TOTAL - Zavala County (La Pryor)		**2,804**	**3.299**	**2,804**	**0**	**0**	**2,804**	**2,804**	**850**	**850**	**850**
EDAP TOTAL ZAVALA COUNTY		**Number of Colonias: 2**	**4,404**	**3.649**	**4,404**	**0**	**0**	**4,404**	**4,404**	**1,207**	**1,207**	**1,207**
Colonias Not Served in EDAP												
		Amaya	196	4.083	196	0	0	196		48	48	48
		Campestre	113	4.520	113	0	0	113		25	25	25
		Camposanto/Elcometa	288	4.000	288	0	0	288		72	72	72
		Chula Vista	444	3.552	444	0	0	444		125	125	318
		La Taverna	100	4.000	100	0	0	100		25	25	25
		Leija	99	4.500	0	99	0	99		22	22	22
		Loma Alta	119	2.479	119	0	0	119		48	48	175
		Nueces Lake	105	3.500	0	105	0	105		30	30	30
		Popeye	35	3.500	0	35	0	35		10	10	10
		Rock Quarry	68	4.533	68	0	0	68		15	15	15
		Surita	25	5.000	25	0	0	25		5	5	5
		Triangula	40	3.636	40	0	0	40		11	11	11
NOT SERVED BY EDAP - TOTAL ZAVALA COUNTY		**Number of Colonias: 12**	**1,632**	**3.743**	**1,393**	**239**	**0**	**1,632**	**0**	**436**	**436**	**756**

Table A.33
Cameron County Cost Summary

Project Status	Project Name / Sponsor	Estimated Total Project Costs	Project Cost per Capita	Project Cost per Capita Served	Project Cost per Dwelling	Project Cost per Dwelling (Density)	Project Cost per Lot at Buildout
Completed Projects	City of Brownsville (Cameron Park)	$6,650,000	$1,512	$1,512	$8,831	$8,831	$4,095
	City of Brownsville (Hacienda Gardens)	$477,800	$1,551	$1,551	$6,545	$6,545	$4,084
Plans and Specifications	Olimito WSC	$7,080,000	$1,868	$1,868	$6,537	$6,537	$6,537
Facilities Planning	Cameron County (Valle Hermosa & Valle Escondido)	$1,011,500	$5,032	$5,032	$24,671	$24,671	$12,335
	Rio Hondo	$536,000	$1,595	$1,595	$6,022	$6,022	$6,022
	Rural Planning/Cameron County	$1,300,000	$418	$418	$1,828	$1,828	$510
	San Benito	$1,031,500	$1,411	$1,411	$6,104	$6,104	$6,032
	Urban Regional Wastewater Planning/Cameron County	$34,250,000	$3,008	$3,008	$14,014	$14,014	$9,030
EDAP TOTAL CAMERON COUNTY		**$52,336,800**	**$2,157**	**$2,157**	**$9,759**	**$9,759**	**$5,504**

Table A.34
Coryell County Cost Summary

Project Status	Project Name / Sponsor	Estimated Total Project Costs	Project Cost per Capita	Project Cost per Capita Served	Project Cost per Dwelling	Project Cost per Dwelling (Density)	Project Cost per Lot at Buildout
Facilities	Copperas Cove	$5,837,500	$3,621	$3,621	$11,140	$11,140	$9,385
Planning	Gatesville	$7,052,188	$8,115	$8,115	$21,833	$21,833	$21,833
EDAP TOTAL CORYELL COUNTY		**$12,889,688**	**$5,195**	**$5,195**	**$15,218**	**$15,218**	**$13,640**

Table A.35
Dimmit County Cost Summary

Project Status	Project Name / Sponsor	Estimated Total Project Costs	Project Cost per Capita	Project Cost per Capita Served	Project Cost per Dwelling	Project Cost per Dwelling (Density)	Project Cost per Lot at Buildout
Facilities Planning	Dimmit County	$2,837,254	$1,201	$2,111	$3,793	$6,666	$617
EDAP TOTAL DIMMIT COUNTY		**$2,837,254**	**$1,201**	**$2,111**	**$3,793**	**$6,666**	**$617**

Table A.36
El Paso County Cost Summary

Project Status	Project Name / Sponsor	Estimated Total Project Costs	Project Cost per Capita	Project Cost per Capita Served	Project Cost per Dwelling	Project Cost per Dwelling (Density)	Project Cost per Lot at Buildout
Completed	EPCLVWDA (Socorro-Bauman Water Project)	$1,820,000	$736	$736	$3,991	$3,991	$3,090
	EPCWCID	$1,510,000	$470	$721	$2,217	$3,403	$1,157
Plans and Specifications	City of El Paso	$5,750,000	Calculations are not made because Westway colonia is already served under the EPCWCID project.				
	EPLVWDA Socorro Phase II	$15,480,000	$1,919	$1,919	$10,389	$10,389	$7,694
	EPCLVWDA Socorro/San Elizario - Phase III	$53,000,000	$3,158	$3,158	$17,086	$17,086	$12,646
	Homestead MUD	$9,410,000	$1,871	$1,871	$8,424	$8,424	$5,621
Facilities Planning	City of El Paso (Canutillo)	$3,184,164	$1,355	$1,355	$6,525	$6,525	$4,601
	Eastside Montana	$10,441,928	$1,425	$1,425	$6,600	$6,600	$3,679
	EPLVWDA (Las Azaleas)	$2,019,300	Calculations are not made because Las Azaleas colonia is already served under the EPCLVWDA Socorro/San Elizario - Phase III project.				
	Tornillo WSC	$3,513,157	$3,101	$3,101	$16,809	$16,809	$14,577
EDAP TOTAL* EL PASO COUNTY		**$98,359,249**	**$2,121**	**$2,173**	**$10,779**	**$11,046**	**$7,263**
		*** Not including City of El Paso and EPLVWDA (Las Azaleas) projects.**					

Table A.37
Hidalgo County Cost Summary

Project Status	Project Name / Sponsor	Estimated Total Project Costs	Project Cost Per Capita	Project Cost Per Capita Served	Project Cost Per Dwelling	Project Cost Per Dwelling (Density)	Project Cost Per Lot at Buildout
Completed	Lull	$1,260,000	$972	$972	$4,375	$4,375	$3,621
Under Construction	City of Edinburg (Faysville)	$9,620,000	$3,063	$3,063	$13,763	$13,763	$10,653
	City of Mission (Granjeno & Madero)	$4,500,000	$3,306	$3,306	$21,327	$21,327	$8,789
Plans and Specifications	City of Alton	$8,440,000	$1,172	$1,172	$5,282	$5,282	$3,078
	City of Pharr (Las Milpas)	$10,170,000	$1,184	$1,184	$5,930	$5,930	$5,913
	City of Weslaco	$9,400,000	$3,290	$3,290	$14,850	$14,850	$10,064
	DeAnda & Saenz	$870,000	$6,591	$6,591	$29,000	$29,000	$24,857
Facilities Planning	South Tower Estates	$786,181	$552	$552	$2,838	$2,838	$2,838
	City of Edinburg (Northwest)	$2,446,286	$881	$881	$3,971	$3,971	$2,494
	Donna	$7,046,725	$1,588	$1,588	$7,242	$7,242	$4,959
	Elsa	$3,537,500	$1,631	$1,631	$7,354	$7,354	$3,896
	HC - Urban Regional WW Planning	$22,950,000	$1,295	$1,295	$5,841	$5,841	$3,585
	Sanchez Ranch	$1,044,390	$1,899	$1,899	$11,352	$11,352	$10,444
	La Joya WSC	$8,534,500	$1,979	$1,979	$8,974	$8,974	$5,107
	McAllen	$6,046,125	$2,203	$2,203	$9,944	$9,944	$6,046
	Mercedes	$6,037,719	$1,519	$1,519	$7,145	$7,145	$5,376
	City of Mission (North Mission)	Cost estimate not available as of April, 1996.					
	Palmview	$7,081,750	$879	$879	$3,963	$3,963	$2,105
	Penitas	$1,840,500	$1,013	$1,013	$4,567	$4,567	$3,466
	San Juan	$15,107,420	$3,067	$3,067	$13,822	$13,822	$9,395
	Hidalgo County - El Paraiso	$1,029,631	$797	$797	$3,588	$3,588	$3,111
EDAP TOTAL* HIDALGO COUNTY:		**$127,748,727**	**$1,581**	**$1,581**	**$7,293**	**$7,293**	**$4,747**
	* Not including City of Mission (North Mission) project. Cost estimate was not available as of April, 1996.						

Table A.38
Hudspeth County Cost Summary

Project Status	Project Name / Sponsor	Estimated Total Project Costs	Project Cost per Capita	Project Cost per Capita Served	Project Cost per Dwelling	Project Cost per Dwelling (Density)	Project Cost per Lot at Buildout
Plans and Specifications	Hudspeth WCID #1	$1,240,000	$1,398	$1,398	$6,294	$6,294	$3,543
EDAP TOTAL HUDSPETH COUNTY		**$1,240,000**	**$1,398**	**$1,398**	**$6,294**	**$6,294**	**$3,543**

Table A.39
Kinney County Cost Summary

Project Status	Project Name / Sponsor	Estimated Total Project Costs	Project Cost per Capita	Project Cost per Capita Served	Project Cost per Dwelling	Project Cost per Dwelling (Density)	Project Cost per Lot at Buildout
Facilities Planning	Spofford	$588,000	$7,259	$7,259	$20,276	$20,276	$20,276
EDAP TOTAL KINNEY COUNTY		**$588,000**	**$7,259**	**$7,259**	**$20,276**	**$20,276**	**$20,276**

Table A.40
La Salle County Cost Summary

Project Status	Project Name / Sponsor	Estimated Total Project Costs	Project Cost per Capita	Project Cost per Capita Served	Project Cost per Dwelling	Project Cost per Dwelling (Density)	Project Cost per Lot at Buildout
Facilities Planning	La Salle County	$1,154,400	$1,203	$1,203	$4,810	$4,810	$1,649
EDAP TOTAL LA SALLE COUNTY		**$1,154,400**	**$1,203**	**$1,203**	**$4,810**	**$4,810**	**$1,649**

Table A.41
Maverick County Cost Summary

Project Status	Project Name / Sponsor	Estimated Total Project Costs	Project Cost per Capita	Project Cost per Capita Served	Project Cost per Dwelling	Project Cost per Dwelling (Density)	Project Cost per Lot at Buildout
Under Construction	City of Eagle Pass	$11,070,000	$1,483	$1,483	$6,673	$6,673	$4,567
Facilities Planning	Quernado	$2,028,125	$3,819	$3,819	$17,188	$17,188	$5,083
EDAP TOTAL MAVERICK COUNTY		**$13,098,125**	**$1,638**	**$1,638**	**$7,371**	**$7,371**	**$4,640**

Table A.42
Presidio County Cost Summary

Project Status	Project Name / Sponsor	Estimated Total Project Costs	Project Cost per Capita	Project Cost per Capita Served	Project Cost per Dwelling	Project Cost per Dwelling Density	Project Cost per Lot at Buildout
Facilities Planning	Ruidosa	$420,250	$7,123	$7,123	$42,025	$42,025	$16,810
EDAP TOTAL PRESIDIO COUNTY		**$420,250**	**$7,123**	**$7,123**	**$42,025**	**$42,025**	**$16,810**

Table A.43
San Patricio County Cost Summary

Project Status	Project Name / Sponsor	Estimated Total Project Costs	Project Cost per Capita	Project Cost per Capita Served	Project Cost per Dwelling	Project Cost per Dwelling Density	Project Cost per Lot at Buildout
Facilities Planning	Ingleside	$1,443,094	Information on Ingleside not available on Texas Water Development Board's needs database as of December, 1995.				
	San Patricio County	$24,389,750	$2,586	$2,586	$7,335	$7,335	$7,324
EDAP TOTAL* SAN PATRICIO COUNTY		**$24,389,750**	**$2,586**	**$2,586**	**$7,335**	**$7,335**	**$7,324**
		* Not including Ingleside project. Colonia population, number of dwellings and density not available.					

Table A.44
Starr County Cost Summary

Project Status	Project Name / Sponsor	Estimated Total Project Costs	Project Cost per Capita	Project Cost per Capita Served	Project Cost per Dwelling	Project Cost per Dwelling Density	Project Cost per Lot at Buildout
Plans and Specifications	Starr Co. WCID #2 (Las Lomas)	$1,150,000	$557	$557	$2,110	$2,110	$1,917
Facilities Planning	Roma	$3,222,500	$607	$607	$2,736	$2,736	$2,253
EDAP TOTAL STARR COUNTY		**$4,372,500**	**$593**	**$593**	**$2,538**	**$2,538**	**$2,154**

Table A.45
Terrell County Cost Summary

Project Status	Project Name / Sponsor	Estimated Total Project Costs	Project Cost per Capita	Project Cost per Capita Served	Project Cost per Dwelling	Project Cost per Dwelling (Density)	Project Cost per Lot at Buildout
Facilities Planning	Terrell County WCID (Sanderson)	$2,851,675	$2,852	$2,852	$4,599	$4,599	$3,587
EDAP TOTAL TERRELL COUNTY		**$2,851,675**	**$2,852**	**$2,852**	**$4,599**	**$4,599**	**$3,587**

Table A.46
Uvalde County Cost Summary

Project Status	Project Name / Sponsor	Estimated Total Project Costs	Project Cost per Capita	Project Cost per Capita Served	Project Cost per Dwelling	Project Cost per Dwelling (Density)	Project Cost per Lot at Buildout
Facilities Planning	Uvalde County	$5,623,605	$2,600	$2,648	$6,850	$6,975	$6,200
EDAP TOTAL UVALDE COUNTY		**$5,623,605**	**$2,600**	**$2,648**	**$6,850**	**$6,975**	**$6,200**

Table A.47
Val Verde County Cost Summary

Project Status	Project Name / Sponsor	Estimated Total Project Costs	Project Cost per Capita	Project Cost per Capita Served	Project Cost per Dwelling	Project Cost per Dwelling (Density)	Project Cost per Lot at Buildout
Under Construction	City of Del Rio (Cienegas Terrace)	$3,510,000	$3,510	$3,510	$17,550	$17,550	$4,517
Plans and Specifications	City of Del Rio (Val Verde Park Estates)	$12,010,000	$26,689	$26,689	$120,100	$120,100	$9,717
EDAP TOTAL VAL VERDE COUNTY		**$15,520,000**	**$10,703**	**$10,703**	**$51,733**	**$51,733**	**$7,710**

Table A.48
Webb County Cost Summary

Project Status	Project Name / Sponsor	Estimated Total Project Costs	Project Cost per Capita	Project Cost per Capita Served	Project Cost per Dwelling	Project Cost per Dwelling Density	Project Cost per Lot at Buildout
Plans and Specifications	Webb County (Larga Vista)	$1,570,000	$2,990	$2,990	$14,952	$14,952	$12,171
Facilities Planning	Webb County (Mines Road & S.H. 359)	$9,435,066	$2,503	$2,585	$11,354	$11,727	$8,044
	Webb County (Southwest Webb)	Cost estimate not available as of April, 1996.					
EDAP TOTAL* WEBB COUNTY		**$11,005,066**	**$2,562**	**$2,636**	**$11,758**	**$12,095**	**$8,452**
		* Not including Webb County (Southwest Webb) project. Cost estimate was not available as of April, 1996.					

Table A.49
Willacy County Cost Summary

Project Status	Project Name / Sponsor	Estimated Total Project Costs	Project Cost per Capita	Project Cost per Capita Served	Project Cost per Dwelling	Project Cost per Dwelling (Density)	Project Cost per Lot at Buildout
Complete	Sebastian WSC (Sebastian)	$3,020,000	$1,586	$1,586	$6,426	$6,426	$4,441
Plans and Specifications	North Alamo WSC (La Sara)	$1,811,000	$2,198	$2,198	$8,791	$8,791	$6,077
EDAP TOTAL WILLACY COUNTY		**$4,831,000**	**$1,771**	**$1,771**	**$7,146**	**$7,146**	**$4,940**

Table A.50
Zapata County Cost Summary

Project Status	Project Name / Sponsor	Estimated Total Project Costs	Project Cost per Capita	Project Cost per Capita Served	Project Cost per Dwelling	Project Cost per Dwelling Density	Project Cost per Lot at Buildout
Facilities Planning	Zapata County	$5,051,000	$2,253	$2,253	$8,561	$8,561	$4,534
EDAP TOTAL ZAPATA COUNTY		**$5,051,000**	**$2,253**	**$2,253**	**$8,561**	**$8,561**	**$4,534**

Table A.51
Zavala County Cost Summary

Project Status	Project Name / Sponsor	Estimated Total Project Costs	Project Cost per Capita	Project Cost per Capita Served	Project Cost per Dwelling	Project Cost per Dwelling Density	Project Cost per Lot at Buildout
Facilities Planning	Zavala County (Batesville)	$1,199,004	$749	$749	$3,359	$3,359	$3,359
	Zavala County (La Pryor)	$2,640,607	$942	$942	$3,107	$3,107	$3,107
EDAP TOTAL ZAVALA COUNTY		**$3,839,611**	**$872**	**$872**	**$3,181**	**$3,181**	**$3,181**

Appendix B. Texas Department of Housing and Community Affairs Colonia Infrastructure Funding

Table Notes and Sources

Column Title	Description	Source
Colonia Name	Name of colonia as denoted by the Texas Department of Housing and Community Affairs	Texas Department of Housing and Community Affairs database as provided to the LBJ School of Public Affairs, University of Texas at Austin.
Fund Used	Texas Department of Housing and Community Affairs fund from which grant money is awarded. There are five possible funds: Colonia Fund (CF), Colonia Fund II (CF-2), Colonia Fund - Planning (CFP), Colonia Fund - Construction (CFC), and Colonia Fund - Demonstration (CFD).	Texas Department of Housing and Community Affairs database as provided to the LBJ School of Public Affairs, University of Texas at Austin.
Grant Amount	Dollar amount of grant funds awarded for project.	Texas Department of Housing and Community Affairs database as provided to the LBJ School of Public Affairs, University of Texas at Austin.
Matching Funds	Dollar amount (can be in-kind) of resources obligated by county for project.	Texas Department of Housing and Community Affairs database as provided to the LBJ School of Public Affairs, University of Texas at Austin.
Start Date	Contract start date.	Texas Department of Housing and Community Affairs database as provided to the LBJ School of Public Affairs, University of Texas at Austin.
Planned Completion Date	Planned contract completion date.	Texas Department of Housing and Community Affairs database as provided to the LBJ School of Public Affairs, University of Texas at Austin.
Proposed Low/ Moderate Income Beneficiaries	Estimated number of residents with low or moderate incomes who will benefit from the project.	Texas Department of Housing and Community Affairs database as provided to the LBJ School of Public Affairs, University of Texas at Austin.
Proposed Total Beneficiaries	Estimated total number of residents who will benefit from the project.	Texas Department of Housing and Community Affairs database as provided to the LBJ School of Public Affairs, University of Texas at Austin.

Tables Notes and Sources continued

Column Title	Description	Source
Proposed % Low/ Moderate Income Beneficiaries	Percentage of beneficiaries estimated to have low or moderate incomes.	Texas Department of Housing and Community Affairs database as provided to the LBJ School of Public Affairs, University of Texas at Austin.
Actual Low/ Moderate Income Beneficiaries	Actual number of residents with low or moderate incomes who have benefitted from a compelted project.	Texas Department of Housing and Community Affairs database as provided to the LBJ School of Public Affairs, University of Texas at Austin.
Actual Total Beneficiaries	Actual total number of residents who have benefitted from a project.	Texas Department of Housing and Community Affairs database as provided to the LBJ School of Public Affairs, University of Texas at Austin.
Actual % Low/ Moderate Income Beneficiaries	Percentage of residents with low or moderate incomes which benefitted from a compelted project.	Texas Department of Housing and Community Affairs database as provided to the LBJ School of Public Affairs, University of Texas at Austin.
Distribution of Project Activities	Types of projects which have been awarded funds from TDHCA for colonia improvement.	Calcuated using information in the "proposed activity" field of the Texas Department of Housing and Community Affairs database. Possible activities defined below as water, sewer, septic tank, planning, platting, street paving, drainage, community center, housing, comprehensive demonstration project or other.
Water	Installation or modification of central water distribution infrastructure or service.	Texas Department of Housing and Community Affairs database as provided to the LBJ School of Public Affairs, University of Texas at Austin, "proposed activity" field.
Sewer	Installation or modification of central wastewater collection infrastructure or service.	Texas Department of Housing and Community Affairs database as provided to the LBJ School of Public Affairs, University of Texas at Austin, "proposed activity" field.
Septic Tank	Installation of a septic tank wastewater collection and treatment system.	Texas Department of Housing and Community Affairs database as provided to the LBJ School of Public Affairs, University of Texas at Austin, "proposed activity" field.
Planning	Planning of projects.	Texas Department of Housing and Community Affairs database as provided to the LBJ School of Public Affairs, University of Texas at Austin, "proposed activity" field.

Table Notes and Sources continued

Column Title	Description	Source
Platting	Proper subdivision and recording of properties with county government.	Texas Department of Housing and Community Affairs database as provided to the LBJ School of Public Affairs, University of Texas at Austin, "proposed activity" field.
Street Paving	Paving of asphalt roads.	Texas Department of Housing and Community Affairs database as provided to the LBJ School of Public Affairs, University of Texas at Austin, "proposed activity" field.
Drainage	Installation or modification of storm sewer infrastructure.	Texas Department of Housing and Community Affairs database as provided to the LBJ School of Public Affairs, University of Texas at Austin, "proposed activity" field.
Community Center	Installation or modification of a community center building.	Texas Department of Housing and Community Affairs database as provided to the LBJ School of Public Affairs, University of Texas at Austin, "proposed activity" field.
Housing	Housing rehabilitation projects.	Texas Department of Housing and Community Affairs database as provided to the LBJ School of Public Affairs, University of Texas at Austin, "proposed activity" field.
Comprehensive Demonstration Project	Community devleopment needs within a coloniB.	Texas Department of Housing and Community Affairs database as provided to the LBJ School of Public Affairs, University of Texas at Austin, "proposed activity" field.
Other	Provision or improvment of public parks, recreation areas, or other eligible activities.	Texas Department of Housing and Community Affairs database as provided to the LBJ School of Public Affairs, University of Texas at Austin, "proposed activity" field.

Table B.1
Aransas County TDHCA Colonia Funding

Colonia Name	Fund Used	Grant Amount	Matching Funds	Start Date	Planned Completion Date	Proposed Low/ Moderate Income Beneficiaries	Proposed Total Beneficiaries	Proposed % Low/ Moderate Income Beneficiaries	Actual Low/ Moderate Income Beneficiaries	Actual Total Benefi-ciaries	Actual % Low/ Moderate Income Beneficiaries
Lanfair Lane	CFC	$231,510	$27,000	7/1/95	6/30/97	92	93	98.9%			
TDHCA Total Aransas County		**$231,510**	**$27,000**			**92**	**93**	**98.9%**			

Table B.2
Aransas County TDHCA Project Activity

	Distribution of Project Activities										
	Water	Sewer	Septic Tank	Planning	Platting	Street Paving	Drainage	Community Center	Housing	Comprehensive Demonstration Project	Other
Number of Colonias Benefiting from Activity	1	1				1					

Table B.3
Bee County TDHCA Colonia Funding

Colonia Name	Fund Used	Grant Amount	Matching Funds	Start Date	Planned Completion Date	Proposed Low/ Moderate Income Beneficiaries	Proposed Total Beneficiaries	Proposed % Low/ Moderate Income Beneficiaries	Actual Low/ Moderate Income Beneficiaries	Actual Total Benefic-iaries	Actual % Low/ Moderate Income Beneficiaries
Tynan	CFC	$365,298	$40,000	7/1/95	6/30/97	252	288	87.5%			
Tynan	CFC	$134,547	$40,000	7/1/95	6/30/97	252	288	87.5%			
TDHCA Total Bee County		**$499,845**	**$40,000**			**252**	**288**	**87.5%**			

Table B.4
Bee County TDHCA Project Activity

	Distribution of Project Activities										
	Water	Sewer	Septic Tank	Planning	Platting	Street Paving	Drainage	Community Center	Housing	Comprehensive Demonstration Project	Other
Number of Colonias Benefiting from Activity	1										

Table B.5
Brooks County TDHCA Colonia Funding

Colonia Name	Fund Used	Grant Amount	Matching Funds	Start Date	Planned Completion Date	Proposed Low/ Moderate Income Beneficiaries	Proposed Total Beneficiaries	Proposed % Low/ Moderate Income Beneficiaries	Actual Low/ Moderate Income Beneficiaries	Actual Total Benefic-iaries	Actual % Low/ Moderate Income Beneficiaries
Airport Road Subdivision	CFC	$424,000	$14,300	4/1/94	3/31/96	198	230	86.1%			
Cantu Addition	CFC	$500,000	$34,000	7/1/95	6/30/97	307	334	91.9%			
La Parrita	CFC	$500,000	$34,000	7/1/95	6/30/97	198	230	86.1%			
La Parrita	CFC	$424,000	$14,300	4/1/94	3/31/96	198	230	86.1%			
La Parrita (Flowella)	CF	$486,700	$52,500	5/1/93	4/30/95	100	107	93.5%	100	107	93.5%
TDHCA Total Brooks County		**$1,410,700**	**$100,800**			**605**	**671**	**90.2%**	**100**	**107**	**93.5%**

Table B.6
Brooks County TDHCA Project Activity

	Distribution of Project Activities										
	Water	**Sewer**	**Septic Tank**	**Planning**	**Platting**	**Street Paving**	**Drainage**	**Community Center**	**Housing**	**Comprehensive Demonstration Project**	**Other**
Number of Colonias Benefiting from Activity	3										

Table B.7
Cameron County TDHCA Colonia Funding

Colonia Name	Fund Used	Grant Amount	Matching Funds	Start Date	Planned Completion Date	Proposed Low/ Moderate Income Beneficiaries	Proposed Total Beneficiaries	Proposed % Low/ Moderate Income Beneficiaries	Actual Low/ Moderate Income Beneficiaries	Actual Total Benefic-iaries	Actual % Low/ Moderate Income Beneficiaries
Cameron Park	CFD	$1,000,000		7/28/94	7/27/96	3,307	3,802	87.0%			
Cameron Park	CFP	$71,600	$0	5/1/93	10/31/94	2,764	3,802	72.7%	2,764	3,802	72.7%
Dakota Mobile Homes	CF-2	$300,000	$0	9/9/92	9/8/94	3,307	3,802	87.0%	199	199	100.0%
Jaime Lake	CF-2	$300,000	$17,736	9/9/92	9/8/94	199	199	100.0%	199	199	100.0%
Las Palmas	CF	$500,000	$0	5/1/93	4/30/95	466	466	100.0%	594	594	100.0%
Leal Subdivision (Leal #2)	CFP	$93,754	$0	4/11/94	12/31/95	339	339	100.0%			
Los Cuates	CFC	$391,255	$24,000	5/1/95	4/30/97	111	118	94.1%			
Sunny Skies	CFP	$93,754	$0	4/11/94	12/31/95	339	339	100.0%			
Villa Pancho	CFC	$500,000	$52,165	4/1/94	3/31/96	235	267	88.0%			
TDHCA Total Cameron County		**$2,856,609**	**$93,901**			**10,294**	**12,329**	**83.5%**	**3,557**	**4,595**	**78.3%**

Table B.8
Cameron County TDHCA Project Activity

	Distribution of Project Activities										
	Water	Sewer	Septic Tank	Planning	Platting	Street Paving	Drainage	Community Center	Housing	Comprehensive Demonstration Project	Other
Number of Colonias Benefiting from Activity		5		1	2	1	1	1	1		

Table B.9
Crockett County TDHCA Colonia Funding

Colonia Name	Fund Used	Grant Amount	Matching Funds	Start Date	Planned Completion Date	Proposed Low/ Moderate Income Beneficiaries	Proposed Total Beneficiaries	Proposed % Low/ Moderate Income Beneficiaries	Actual Low/ Moderate Income Beneficiaries	Actual Total Benefic-iaries	Actual % Low/ Moderate Income Beneficiaries
Ozona	CFC	$495,800	$45,000	7/1/95	6/30/97	412	428	96.3%			
TDHCA Total Crockett County		**$495,800**	**$45,000**			**412**	**428**	**96.3%**			

Table B.10
Crockett County TDHCA Project Activity

	Distribution of Project Activities										
	Water	Sewer	Septic Tank	Planning	Platting	Street Paving	Drainage	Community Center	Housing	Comprehensive Demonstration Project	Other
Number of Colonias Benefiting from Activity	1	1									

Table B.11
Dimmit County TDHCA Colonia Funding

Colonia Name	Fund Used	Grant Amount	Matching Funds	Start Date	Planned Completion Date	Proposed Low/ Moderate Income Beneficiaries	Proposed Total Beneficiaries	Proposed % Low/ Moderate Income Beneficiaries	Actual Low/ Moderate Income Beneficiaries	Actual Total Benefic-iaries	Actual % Low/ Moderate Income Beneficiaries
Brundage	CFC	$297,825	$15,675	5/1/95	4/30/97	34	39	87.2%			
Catarina	CFP	$19,750	$0	5/1/95	4/30/97	153	172	89.0%			
TDHCA Total Dimmit County		**$317,575**	**$15,675**			**187**	**211**	**88.6%**			

Table B.12
Dimmit County TDHCA Project Activity

	Distribution of Project Activities										
	Water	Sewer	Septic Tank	Planning	Platting	Street Paving	Drainage	Community Center	Housing	Comprehensive Demonstration Project	Other
Number of Colonias Benefiting from Activity	1			1							

Table B.13
Duval County TDHCA Colonia Funding

Colonia Name	Fund Used	Grant Amount	Matching Funds	Start Date	Planned Completion Date	Proposed Low/ Moderate Income Beneficiaries	Proposed Total Beneficiaries	Proposed % Low/ Moderate Income Beneficiaries	Actual Low/ Moderate Income Beneficiaries	Actual Total Benefic-iaries	Actual % Low/ Moderate Income Beneficiaries
Buena Vista	CF-2	$297,890	$21,801	9/9/92	9/8/94	180	200	90.0%	277	325	85.2%
Cadena	CF-2	$297,890	$21,801	9/9/92	9/8/94	180	200	90.0%	277	325	85.2%
TDHCA Total Duval County		**$297,890**	**$21,801**			**180**	**200**	**90.0%**	**277**	**325**	**85.2%**

Table B.14
Duval County TDHCA Project Activity

	Distribution of Project Activities										
	Water	Sewer	Septic Tank	Planning	Platting	Street Paving	Drainage	Community Center	Housing	Comprehensive Demonstration Project	Other
Number of Colonias Benefiting from Activity	2										

Table B.15
Ector County TDHCA Colonia Funding

Colonia Name	Fund Used	Grant Amount	Matching Funds	Start Date	Planned Completion Date	Proposed Low/ Moderate Income Beneficiaries	Proposed Total Beneficiaries	Proposed % Low/ Moderate Income Beneficiaries	Actual Low/ Moderate Income Beneficiaries	Actual Total Benefic- iaries	Actual % Low/ Moderate Income Beneficiaries
West Odessa	CFP	$27,500	$0	5/1/95	4/30/97	4,340	7,551	57.5%			
TDHCA Total Ector County		**$27,500**	**$0**			**4,340**	**7,551**	**57.5%**			

Table B.16
Ector County TDHCA Project Activity

	Distribution of Project Activities										
	Water	Sewer	Septic Tank	Planning	Platting	Street Paving	Drainage	Community Center	Housing	Comprehensive Demonstration Project	Other
Number of Colonias Benefiting from Activity				1							

Table B.17
El Paso County TDHCA Colonia Funding

Colonia Name	Fund Used	Grant Amount	Matching Funds	Start Date	Planned Completion Date	Proposed Low/ Moderate Income Beneficiaries	Proposed Total Beneficiaries	Proposed % Low/ Moderate Income Beneficiaries	Actual Low/ Moderate Income Beneficiaries	Actual Total Benefic-iaries	Actual % Low/ Moderate Income Beneficiaries
Bosque Bonito	CFC	$500,000	$50,000	5/1/95	4/30/97	758	770	98.4%			
Brinkman Addition	CFC	$500,000	$50,000	4/1/94	3/31/96	575	591	97.3%	526	555	94.8%
Camino Barrial	CFC	$500,000	$50,000	4/1/94	3/31/96	758	770	98.4%	526	555	94.8%
Gloria Elena	CFC	$500,000	$50,000	4/1/94	3/31/96	575	591	97.3%	526	555	94.8%
Gonzalez	CFC	$500,000	$50,000	4/1/94	3/31/96	575	591	97.3%	526	555	94.8%
Sparks	CFD	$1,000,000	$4,151,124	4/1/94	3/31/97	1,095	1,276	85.8%			
TDHCA Total El Paso County		**$2,000,000**	**$4,251,124**			**2,428**	**2,637**	**92.1%**	**526**	**555**	**94.8%**

Table B.18
El Paso County TDHCA Project Activity

	Distribution of Project Activities										
	Water	**Sewer**	**Septic Tank**	**Planning**	**Platting**	**Street Paving**	**Drainage**	**Community Center**	**Housing**	**Comprehensive Demonstration Project**	**Other**
Number of Colonias Benefiting from Activity	5									1	

Table B.19
Frio County TDHCA Colonia Funding

Colonia Name	Fund Used	Grant Amount	Matching Funds	Start Date	Planned Completion Date	Proposed Low/ Moderate Income Beneficiaries	Proposed Total Beneficiaries	Proposed % Low/ Moderate Income Beneficiaries	Actual Low/ Moderate Income Beneficiaries	Actual Total Beneflc-iaries	Actual % Low/ Moderate Income Beneficiaries
Big Foot	CF-2	$300,000	$176,000	9/1/92	8/31/94	180	180	100.0%	318	388	82.0%
Alta Vista & Frio Heights	CFP	$36,500	$8,000	4/11/94	12/31/95	555	645	86.0%	555	645	86.0%
Big Foot	CFP	$36,500	$8,000	4/11/94	12/31/95	180	180	100.0%	555	645	86.0%
Hilltop	CFP	$36,500	$8,000	4/11/94	12/31/95	555	645	86.0%	555	645	86.0%
Alta Vista & Frio Heights	CFC	$300,000	$20,000	5/1/95	4/30/97	91	91	100.0%			
Big Foot	CFC	$300,000	$20,000	5/1/95	4/30/97	91	91	100.0%			
Hilltop	CFC	$300,000	$20,000	5/1/95	4/30/97	91	91	100.0%			
TDHCA Total Frio County		**$636,500**	**$204,000**			**826**	**916**	**90.2%**	**873**	**1,033**	**84.5%**

Table B.20
Frio County TDHCA Project Activity

	Distribution of Project Activities										
	Water	Sewer	Septic Tank	Planning	Platting	Street Paving	Drainage	Community Center	Housing	Comprehensive Demonstration Project	Other
Number of Colonias Benefiting from Activity	1	1		3					3		

Table B.21
Gillespie County TDHCA Colonia Funding

Colonia Name	Fund Used	Grant Amount	Matching Funds	Start Date	Planned Completion Date	Proposed Low/ Moderate Income Beneficiaries	Proposed Total Beneficiaries	Proposed % Low/ Moderate Income Beneficiaries	Actual Low/ Moderate Income Beneficiaries	Actual Total Benefic-iaries	Actual % Low/ Moderate Income Beneficiaries
Colonia Stonewall Community	CF-2	$300,000	$60,000	9/1/92	8/31/94	169	169	100.0%	169	169	100.0%
TDHCA Total Gillespie County		**$300,000**	**$60,000**			**169**	**169**	**100.0%**	**169**	**169**	**100.0%**

Table B.22
Gillespie County TDHCA Project Activity

	Distribution of Project Activities										
	Water	Sewer	Septic Tank	Planning	Platting	Street Paving	Drainage	Community Center	Housing	Comprehensive Demonstration Project	Other
Number of Colonias Benefiting from Activity	1										

Table B.23
Glasscock County TDHCA Colonia Funding

Colonia Name	Fund Used	Grant Amount	Matching Funds	Start Date	Planned Completion Date	Proposed Low/ Moderate Income Beneficiaries	Proposed Total Beneficiaries	Proposed % Low/ Moderate Income Beneficiaries	Actual Low/ Moderate Income Beneficiaries	Actual Total Benefic-iaries	Actual % Low/ Moderate Income Beneficiaries
Garden City	CFP	$42,250	$0	4/11/94	12/31/95	150	273	54.9%			
Garden City	CFC	$500,000	$25,093	5/1/95	4/30/97	50	56	89.3%			
TDHCA Total Glasscock County		**$542,250**	**$25,093**			**200**	**329**	**60.8%**			

Table B.24
Glasscock County TDHCA Project Activity

	Distribution of Project Activities										
	Water	Sewer	Septic Tank	Planning	Platting	Street Paving	Drainage	Community Center	Housing	Comprehensive Demonstration Project	Other
Number of Colonias Benefiting from Activity	1			1							

Table B.25
Hidalgo County TDHCA Colonia Funding

Colonia Name	Fund Used	Grant Amount	Matching Funds	Start Date	Planned Completion Date	Proposed Low/ Moderate Income Beneficiaries	Proposed Total Beneficiaries	Proposed % Low/ Moderate Income Beneficiaries	Actual Low/ Moderate Income Beneficiaries	Actual Total Benefic-iaries	Actual % Low/ Moderate Income Benefic-iaries
Monte Alto	CFP	$37,850	$0	3/4/92	3/3/94	878	1,376	63.8%	878	1,376	63.8%
Bar 8	CF	$500,000	$0	3/4/92	3/3/94	893	893	100.0%	626	893	70.1%
Bertha	CF	$500,000	$0	3/4/92	3/3/94	878	1,376	63.8%	626	893	70.1%
Catherine	CF	$500,000	$0	3/4/92	3/3/94	893	893	100.0%	626	893	70.1%
Colonia #1	CF	$500,000	$0	3/4/92	3/3/94	893	893	100.0%	626	893	70.1%
Colonia #2	CF	$500,000	$0	3/4/92	3/3/94	893	893	100.0%	626	893	70.1%
Country Village #2	CF	$500,000	$0	3/4/92	3/3/94	893	893	100.0%	626	893	70.1%
Old Rebel Heights 1 & 2	CF	$500,000	$0	3/4/92	3/3/94	893	893	100.0%	626	893	70.1%
Palm Lake #1 - 4	CF	$500,000	$0	3/4/92	3/3/94	893	893	100.0%	626	893	70.1%
Seminary Est.	CF	$500,000	$0	3/4/92	3/3/94	893	893	100.0%	626	893	70.1%
Tierra Maria	CF	$500,000	$0	3/4/92	3/3/94	893	893	100.0%	626	893	70.1%
Tom Gill Road	CF	$500,000	$0	3/4/92	3/3/94	893	893	100.0%	626	893	70.1%
El Charro	CF-2	$300,000	$52,500	9/1/92	8/31/94	805	823	97.8%	774	808	95.8%
Tom Weekley	CF-2	$300,000	$52,500	9/1/92	8/31/94	805	823	97.8%	774	808	95.8%
Angela	CF	$440,615	$117,240	5/1/93	4/30/95	912	942	96.8%	1,164	1,194	97.5%
Good Valley Ranch	CF	$440,615	$117,240	5/1/93	4/30/95	912	942	96.8%	1,164	1,194	97.5%
H. M. E.	CF	$440,615	$117,240	5/1/93	4/30/95	912	942	96.8%	1,164	1,194	97.5%
Sing	CF	$440,615	$117,240	5/1/93	4/30/95	912	942	96.8%	1,164	1,194	97.5%
St Louis Groves	CF	$440,615	$117,240	5/1/93	4/30/95	912	942	96.8%	1,164	1,194	97.5%
Victoria Belen	CF	$440,615	$117,240	5/1/93	4/30/95	912	942	96.8%	1,164	1,194	97.5%
Monte Alto	CFC	$500,000	$532,700	4/1/94	3/31/96	1,135	1,271	89.3%	1,135	1,271	89.3%
Boyce	CFP	$61,550	$0	5/1/95	9/30/96	411	432	95.1%			
Chapa	CFP	$61,550	$0	5/1/95	9/30/96	411	432	95.1%			
Flea Market Subd	CFP	$61,550	$0	5/1/95	9/30/96	411	432	95.1%			
Olivarez #3	CFP	$61,550	$0	5/1/95	9/30/96	411	432	95.1%			
Esperanza Est	CFC	$498,523	$269,811	5/1/95	4/30/97	1,268	1,396	90.8%			
Villa Del Mundo	CFC	$498,523	$269,811	5/1/95	4/30/97	1,268	1,396	90.8%			
Villa Del Sol	CFC	$498,523	$269,811	5/1/95	4/30/97	1,268	1,396	90.8%			
TDHCA Total Hidalgo County		**$2,338,538**	**$972,251**			**6,302**	**7,133**	**88.3%**	**4,577**	**5,542**	**82.6%**

Table B.26
Hidalgo County TDHCA Project Activity

	Distribution of Project Activities										
	Water	Sewer	Septic Tank	Planning	Platting	Street Paving	Drainage	Community Center	Housing	Comprehensive Demonstration Project	Other
Number of Colonias Benefiting from Activity	21	11		1	4						

Table B.27
Hudspeth County TDHCA Colonia Funding

Colonia Name	Fund Used	Grant Amount	Matching Funds	Start Date	Planned Completion Date	Proposed Low/ Moderate Income Beneficiaries	Proposed Total Beneficiaries	Proposed % Low/ Moderate Income Beneficiaries	Actual Low/ Moderate Income Beneficiaries	Actual Total Benefic-iaries	Actual % Low/ Moderate Income Beneficiaries
Acala	CFP	$20,000	$1,208	5/1/93	12/31/94	71	73	97.3%	71	73	97.3%
TDHCA Total Hudspeth County		**$20,000**	**$1,208**			**71**	**73**	**97.3%**	**71**	**73**	**97.3%**

Table B.28
Hudspeth County TDHCA Project Activity

	Distribution of Project Activities										
	Water	Sewer	Septic Tank	Planning	Platting	Street Paving	Drainage	Community Center	Housing	Comprehensive Demonstration Project	Other
Number of Colonias Benefiting from Activity				1							

Table B.29
Jim Wells County TDHCA Colonia Funding

Colonia Name	Fund Used	Grant Amount	Matching Funds	Start Date	Planned Completion Date	Proposed Low/ Moderate Income Beneficiaries	Proposed Total Beneficiaries	Proposed % Low/ Moderate Income Beneficiaries	Actual Low/ Moderate Income Beneficiaries	Actual Total Benefic-iaries	Actual % Low/ Moderate Income Beneficiaries
Alice Acres	CFP	$30,750	$0	4/11/94	12/31/95	357	451	79.2%			
Coyote Acres	CFP	$30,750	$0	4/11/94	12/31/95	357	451	79.2%			
English Acres	CFP	$30,750	$0	4/11/94	12/31/95	357	451	79.2%			
K-Bar	CFP	$30,750	$0	4/11/94	12/31/95	357	451	79.2%			
Rancho Alegre	CFC	$500,000	$33,000	4/1/94	3/31/96	189	210	90.0%	189	210	90.0%
TDHCA Total Jim Wells County		**$530,750**	**$33,000**			**546**	**661**	**82.6%**	**189**	**210**	**90.0%**

Table B.30
Jim Wells County TDHCA Project Activity

	Distribution of Project Activities										
	Water	Sewer	Septic Tank	Planning	Platting	Street Paving	Drainage	Community Center	Housing	Comprehensive Demonstration Project	Other
Number of Colonias Benefiting from Activity		1	4								

Table B.31
Karnes County TDHCA Colonia Funding

Colonia Name	Fund Used	Grant Amount	Matching Funds	Start Date	Planned Completion Date	Proposed Low/ Moderate Income Beneficiaries	Proposed Total Beneficiaries	Proposed % Low/ Moderate Income Beneficiaries	Actual Low/ Moderate Income Beneficiaries	Actual Total Benefic-iaries	Actual % Low/ Moderate Income Beneficiaries
Choate	CFP	$36,500	$0	5/1/95	4/30/97	301	345	87.2%			
Helena	CFP	$36,500	$0	5/1/95	4/30/97	301	345	87.2%			
Hobson	CFP	$36,500	$0	5/1/95	4/30/97	301	345	87.2%			
Choate	CFC	$500,000	$0	7/1/95	6/30/97	273	273	100.0%			
Helena	CFC	$500,000	$0	7/1/95	6/30/97	273	273	100.0%			
Hobson	CFC	$500,000	$0	7/1/95	6/30/97	273	273	100.0%			
TDHCA Total Karnes County		**$536,500**	**$0**			**574**	**618**	**92.9%**			

Table B.32
Karnes County TDHCA Project Activity

	Distribution of Project Activities										
	Water	Sewer	Septic Tank	Planning	Platting	Street Paving	Drainage	Community Center	Housing	Comprehensive Demonstration Project	Other
Number of Colonias Benefiting from Activity			3	3							

Table B.33
Kleberg County TDHCA Colonia Funding

Colonia Name	Fund Used	Grant Amount	Matching Funds	Start Date	Planned Completion Date	Proposed Low/ Moderate Income Beneficiaries	Proposed Total Beneficiaries	Proposed % Low/ Moderate Income Beneficiaries	Actual Low/ Moderate Income Beneficiaries	Actual Total Benefic-iaries	Actual % Low/ Moderate Income Beneficiaries
West Riviera											
Ricardo	CFC	$278,320	$8,400	6/1/94	5/31/96	83	103	80.6%			
Ricardo	CFC	$488,600	$6,600	7/1/95	6/30/97	436	525	83.0%			
Riviera	CFC	$488,600	$6,600	7/1/95	6/30/97	436	525	83.0%			
TDHCA Total Kleberg County		**$766,920**	**$15,000**			**519**	**628**	**82.6%**			

Table B.34
Kleberg County TDHCA Project Activity

	Distribution of Project Activities										
	Water	Sewer	Septic Tank	Planning	Platting	Street Paving	Drainage	Community Center	Housing	Comprehensive Demonstration Project	Other
Number of Colonias Benefiting from Activity		2									

Table B.35
La Salle County TDHCA Colonia Funding

Colonia Name	Fund Used	Grant Amount	Matching Funds	Start Date	Planned Completion Date	Proposed Low/ Moderate Income Beneficiaries	Proposed Total Beneficiaries	Proposed % Low/ Moderate Income Beneficiaries	Actual Low/ Moderate Income Beneficiaries	Actual Total Benefic-iaries	Actual % Low/ Moderate Income Beneficiaries
Artesia Wells	CFP	$31,500	$0	5/1/95	4/30/97	101	111	91.0%			
Gardendale	CFP	$31,500	$0	5/1/95	4/30/97	101	111	91.0%			
Los Angeles	CFP	$31,500	$0	5/1/95	4/30/97	101	111	91.0%			
Millet	CFP	$31,500	$0	5/1/95	4/30/97	101	111	91.0%			
Fowlerton	CFC	$475,000	$25,000	5/1/95	4/30/97	100	102	98.0%			
TDHCA Total La Salle County		**$506,500**	**$25,000**			**201**	**213**	**94.4%**			

Table B.36
La Salle County TDHCA Project Activity

	Distribution of Project Activities										
	Water	Sewer	Septic Tank	Planning	Platting	Street Paving	Drainage	Community Center	Housing	Comprehensive Demonstration Project	Other
Number of Colonias Benefiting from Activity	1			4							

Table B.37
Live Oak County TDHCA Colonia Funding

Colonia Name	Fund Used	Grant Amount	Matching Funds	Start Date	Planned Completion Date	Proposed Low/ Moderate Income Beneficiaries	Proposed Total Beneficiaries	Proposed % Low/ Moderate Income Beneficiaries	Actual Low/ Moderate Income Beneficiaries	Actual Total Benefic-iaries	Actual % Low/ Moderate Income Beneficiaries
Lopezville	CF	$396,000	$20,000	5/1/93	4/30/95	141	145	97.2%	160	211	75.8%
TDHCA Total Live Oak County		**$396,000**	**$20,000**			**141**	**145**	**97.2%**	**160**	**211**	**75.8%**

Table B.38
Live Oak County TDHCA Project Activity

	Distribution of Project Activities										
	Water	Sewer	Septic Tank	Planning	Platting	Street Paving	Drainage	Community Center	Housing	Comprehensive Demonstration Project	Other
Number of Colonias Benefiting from Activity		1									

Table B.39
Maverick County TDHCA Colonia Funding

Colonia Name	Fund Used	Grant Amount	Matching Funds	Start Date	Planned Completion Date	Proposed Low/ Moderate Income Beneficiaries	Proposed Total Beneficiaries	Proposed % Low/ Moderate Income Beneficiaries	Actual Low/ Moderate Income Beneficiaries	Actual Total Benefic-iaries	Actual % Low/ Moderate Income Beneficiaries
Colonia Seco Mines	CF	$52,481	$0	3/4/92	3/3/94	199	199	100.0%	199	199	100.0%
Siesta Acres	CF	$52,481	$0	3/4/92	3/3/94	199	199	100.0%	199	199	100.0%
Quemado	CFP	$73,500	$0	5/1/93	10/31/94	726	1,351	53.7%	726	1351	53.7%
Radar Base	CFP	$73,500	$0	5/1/93	10/31/94	726	1,351	53.7%	726	1351	53.7%
Las Brisas I,II, &III	CFC	$500,000	$0	4/1/94	3/31/96	971	986	98.5%			
Las Quintas Fronterizas	CFP	$100,000	$0	4/1/94	12/31/95	876	879	99.7%			
Fabrica Townsite Colonias.	CFD	$1,000,000	$0	7/28/94	7/27/96	1,030	1,092	94.3%			
Seco Mines	CFD	$1,000,000	$0	7/28/94	7/27/96	1,030	1,092	94.3%			
Las Brisas	CFP	$100,000	$0	5/1/95	4/30/97	1,007	1,013	99.4%			
Las Quintas Fronterizas	CFP	$100,000	$0	5/1/95	4/30/97	1,007	1,013	99.4%			
Nellis Lands	CFP	$100,000	$0	5/1/95	4/30/97	1,007	1,013	99.4%			
Eagle Heights	CFC	$500,000	$0	5/1/95	4/30/97	769	783	98.2%			
Lago Vista	CFC	$500,000	$0	5/1/95	4/30/97	769	783	98.2%			
Las Quintas Fronterizas	CFC	$500,000	$0	5/1/95	4/30/97	769	783	98.2%			
Loma Bonita	CFC	$500,000	$0	5/1/95	4/30/97	769	783	98.2%			
TDHCA Total Maverick County		**$2,325,981**	**$0**			**5,578**	**6,303**	**88.5%**	**925**	**1,550**	**59.7%**

Table B.40
Maverick County TDHCA Project Activity

	Distribution of Project Activities										
	Water	Sewer	Septic Tank	Planning	Platting	Street Paving	Drainage	Community Center	Housing	Comprehensive Demonstration Project	Other
Number of Colonias Benefiting from Activity	6	7		2	4	2	2	2	2		

Table B.41
Nueces County TDHCA Colonia Funding

Colonia Name	Fund Used	Grant Amount	Matching Funds	Start Date	Planned Completion Date	Proposed Low/ Moderate Income Beneficiaries	Proposed Total Beneficiaries	Proposed % Low/ Moderate Income Beneficiaries	Actual Low/ Moderate Income Beneficiaries	Actual Total Benefic-iaries	Actual % Low/ Moderate Income Beneficiaries
Banquete	CFP	$53,750	$0	4/11/94	12/31/95	1,043	1,539	67.8%			
Rancho Banquete	CFP	$53,750	$0	4/11/94	12/31/95	1,043	1,539	67.8%			
Spring Gardens	CFP	$53,750	$0	4/11/94	12/31/95	1,043	1,539	67.8%			
Rancho Banquete	CFC	$335,000	$195,000	5/1/95	4/30/97	666	825	80.7%			
Spring Gardens	CFC	$335,000	$195,000	5/1/95	4/30/97	666	825	80.7%			
TDHCA Total Nueces County		**$388,750**	**$195,000**			**1,709**	**2,364**	**72.3%**			

Table B.42
Nueces County TDHCA Project Activity

	Distribution of Project Activities										
	Water	Sewer	Septic Tank	Planning	Platting	Street Paving	Drainage	Community Center	Housing	Comprehensive Demonstration Project	Other
Number of Colonias Benefiting from Activity	2			3							

Table B.43
Pecos County TDHCA Colonia Funding

Colonia Name	Fund Used	Grant Amount	Matching Funds	Start Date	Planned Completion Date	Proposed Low/ Moderate Income Beneficiaries	Proposed Total Beneficiaries	Proposed % Low/ Moderate Income Beneficiaries	Actual Low/ Moderate Income Beneficiaries	Actual Total Benefic-iaries	Actual % Low/ Moderate Income Beneficiaries
Alamo Ranchettes	CFC	$500,000	$25,000	7/1/95	6/30/97	175	218	80.3%			
Belanger	CF-2	$299,999	$16,973	9/1/92	8/31/94	208	236	88.1%	226	254	89.0%
TDHCA Total Pecos County		**$799,999**	**$41,973**			**383**	**454**	**84.4%**	**226**	**254**	**89.0%**

Table B.44
Pecos County TDHCA Project Activity

	Distribution of Project Activities										
	Water	**Sewer**	**Septic Tank**	**Planning**	**Platting**	**Street Paving**	**Drainage**	**Community Center**	**Housing**	**Comprehensive Demonstration Project**	**Other**
Number of Colonias Benefiting from Activity	1	1									

Table B.45
Presidio County TDHCA Colonia Funding

Colonia Name	Fund Used	Grant Amount	Matching Funds	Start Date	Planned Completion Date	Proposed Low/ Moderate Income Beneficiaries	Proposed Total Beneficiaries	Proposed % Low/ Moderate Income Beneficiaries	Actual Low/ Moderate Income Beneficiaries	Actual Total Benefic-iaries	Actual % Low/ Moderate Income Beneficiaries
Redford	CF	$500,000	$0	3/4/92	3/3/94	151	151	100.0%	220	220	100.0%
Ruidosa	CF-2	$300,000	$15,000	9/1/92	8/31/94	56	59	94.9%			
Redford	CF	$500,000	$72,200	5/1/93	4/30/95	151	151	100.0%			
Candelaria	CFC	$500,000	$25,000	7/1/95	6/30/97	198	205	96.6%			
Ruidosa	CFC	$500,000	$25,000	7/1/95	6/30/97	198	205	96.6%			
TDHCA Total Presidio County		**$1,800,000**	**$112,200**			**358**	**361**	**99.2%**	**220**	**220**	**100.0%**

Table B.46
Presidio County TDHCA Project Activity

	Distribution of Project Activities										
	Water	Sewer	Septic Tank	Planning	Platting	Street Paving	Drainage	Community Center	Housing	Comprehensive Demonstration Project	Other
Number of Colonias Benefiting from Activity	2	3							2		

Table B.47
Real County TDHCA Colonia Funding

Colonia Name	Fund Used	Grant Amount	Matching Funds	Start Date	Planned Completion Date	Proposed Low/ Moderate Income Beneficiaries	Proposed Total Beneficiaries	Proposed % Low/ Moderate Income Beneficiaries	Actual Low/ Moderate Income Beneficiaries	Actual Total Benefic-iaries	Actual % Low/ Moderate Income Beneficiaries
Oakmont Village	CFC	$483,940	$25,844	7/1/95	6/30/97	237	270	87.8%			
Saddle Mountain	CFC	$483,940	$25,844	7/1/95	6/30/97	237	270	87.8%			
TDHCA Total Real County		**$483,940**	**$25,844**			**237**	**270**	**87.8%**			

Table B.48
Real County TDHCA Project Activity

	Distribution of Project Activities										
	Water	Sewer	Septic Tank	Planning	Platting	Street Paving	Drainage	Community Center	Housing	Comprehensive Demonstration Project	Other
Number of Colonias Benefiting from Activity	2					2					

Table B.49
Reeves County TDHCA Colonia Funding

Colonia Name	Fund Used	Grant Amount	Matching Funds	Start Date	Planned Completion Date	Proposed Low/ Moderate Income Beneficiaries	Proposed Total Beneficiaries	Proposed % Low/ Moderate Income Beneficiaries	Actual Low/ Moderate Income Beneficiaries	Actual Total Benefic-iaries	Actual % Low/ Moderate Income Beneficiaries
Lindsay Division [TDHCA calls it Green Acres]	CF-2	$300,000	$15,000	9/9/92	9/8/94	269	698	38.5%	270	270	100.0%
Lindsay Division [TDHCA calls it Green Acres]	CF	$375,000	$15,000	5/1/93	4/30/95	414	414	100.0%	285	285	100.0%
Brogado	CFC	$500,000	$25,000	7/1/95	6/30/97	233	233	100.0%			
Orla	CFC	$500,000	$25,000	7/1/95	6/30/97	233	233	100.0%			
Patrole	CFC	$500,000	$25,000	7/1/95	6/30/97	233	233	100.0%			
Red Bluff	CFC	$500,000	$25,000	7/1/95	6/30/97	233	233	100.0%			
Saragosa	CFC	$500,000	$25,000	7/1/95	6/30/97	233	233	100.0%			
TDHCA Total Reeves County		**$1,175,000**	**$55,000**			**916**	**1,345**	**68.1%**	**555**	**555**	**100.0%**

Table B.50
Reeves County TDHCA Project Activity

	Distribution of Project Activities										
	Water	Sewer	Septic Tank	Planning	Platting	Street Paving	Drainage	Community Center	Housing	Comprehensive Demonstration Project	Other
Number of Colonias Benefiting from Activity			7								

Table B.51
San Patricio County TDHCA Colonia Funding

Colonia Name	Fund Used	Grant Amount	Matching Funds	Start Date	Planned Completion Date	Proposed Low/ Moderate Income Beneficiaries	Proposed Total Beneficiaries	Proposed % Low/ Moderate Income Beneficiaries	Actual Low/ Moderate Income Beneficiaries	Actual Total Benefic-iaries	Actual % Low/ Moderate Income Beneficiaries
Edroy	CF	$427,500	$0	3/4/92	3/3/94	321	321	100.0%	321	340	94.4%
J. G. Gonzales	CF-2	$300,000	$70,000	9/9/92	9/8/94	236	249	94.8%	349	369	94.6%
Hidalgo	CF	$444,800	$30,000	5/1/93	4/30/95	763	767	99.5%	860	865	99.4%
Hidalgo 3,4,& 5	CFC	$439,500	$15,000	5/1/95	4/30/97	763	767	99.5%			
TDHCA Total San Patricio County		**$1,611,800**	**$115,000**			**2,083**	**2,104**	**99.0%**	**1,530**	**1,574**	**97.2%**

Table B.52
San Patricio County TDHCA Project Activity

	Distribution of Project Activities										
	Water	Sewer	Septic Tank	Planning	Platting	Street Paving	Drainage	Community Center	Housing	Comprehensive Demonstration Project	Other
Number of Colonias Benefiting from Activity	1	3									

Table B.53
Starr County TDHCA Colonia Funding

Colonia Name	Fund Used	Grant Amount	Matching Funds	Start Date	Planned Completion Date	Proposed Low/ Moderate Income Beneficiaries	Proposed Total Beneficiaries	Proposed % Low/ Moderate Income Beneficiaries	Actual Low/ Moderate Income Beneficiaries	Actual Total Benefic-iaries	Actual % Low/ Moderate Income Beneficiaries
Zarate	CF	$281,915	$0	3/4/92	3/3/94	75	75	100.0%	63	63	100.0%
Salineno	CFC	$500,000	$0	4/1/94	3/31/96	594	630	94.3%			
De La Garza Subdivision	CFC	$500,000	$0	5/1/95	4/30/97	126	134	94.0%			
Olivia Guterrez	CFC	$500,000	$0	5/1/95	4/30/97	126	134	94.0%			
Ranchitos del Norte	CFC	$500,000	$0	5/1/95	4/30/97	126	134	94.0%			
West Alto Bonito	CFC	$500,000	$0	5/1/95	4/30/97	126	134	94.0%			
TDHCA Total Starr County		**$1,281,915**	**$0**			**795**	**839**	**94.8%**	**63**	**63**	**100.0%**

Table B.54
Starr County TDHCA Project Activity

	Distribution of Project Activities										
	Water	Sewer	Septic Tank	Planning	Platting	Street Paving	Drainage	Community Center	Housing	Comprehensive Demonstration Project	Other
Number of Colonias Benefiting from Activity	5	1							4		

Table B.55
Terrell County TDHCA Colonia Funding

Colonia Name	Fund Used	Grant Amount	Matching Funds	Start Date	Planned Completion Date	Proposed Low/ Moderate Income Beneficiaries	Proposed Total Beneficiaries	Proposed % Low/ Moderate Income Beneficiaries	Actual Low/ Moderate Income Beneficiaries	Actual Total Benefic-iaries	Actual % Low/ Moderate Income Beneficiaries
Sanderson	CFP	$56,750	$0	5/1/93	12/31/94	613	1,136	54.0%	613	1,136	54.0%
TDHCA Total Terrell County		**$56,750**	**$0**			**613**	**1,136**	**54.0%**	**613**	**1,136**	**54.0%**

Table B.56
Terrell County TDHCA Project Activity

	Distribution of Project Activities										
	Water	Sewer	Septic Tank	Planning	Platting	Street Paving	Drainage	Community Center	Housing	Comprehensive Demonstration Project	Other
Number of Colonias Benefiting from Activity				1							

Table B.57
Tom Green County TDHCA Colonia Funding

Colonia Name	Fund Used	Grant Amount	Matching Funds	Start Date	Planned Completion Date	Proposed Low/ Moderate Income Beneficiaries	Proposed Total Beneficiaries	Proposed % Low/ Moderate Income Beneficiaries	Actual Low/ Moderate Income Beneficiaries	Actual Total Benefic-iaries	Actual % Low/ Moderate Income Beneficiaries
Red Creek	CF	$425,881	$710,000	7/20/93	7/19/95	292	522	55.9%			
TDHCA Total Tom Green County		**$425,881**	**$710,000**			**292**	**522**				

Table B.58
Tom Green County TDHCA Project Activity

	Distribution of Project Activities										
	Water	Sewer	Septic Tank	Planning	Platting	Street Paving	Drainage	Community Center	Housing	Comprehensive Demonstration Project	Other
Number of Colonias Benefiting from Activity	1										

Table B.59
Uvalde County TDHCA Colonia Funding

Colonia Name	Fund Used	Grant Amount	Matching Funds	Start Date	Planned Completion Date	Proposed Low/ Moderate Income Beneficiaries	Proposed Total Beneficiaries	Proposed % Low/ Moderate Income Beneficiaries	Actual Low/ Moderate Income Beneficiaries	Actual Total Benefic- iaries	Actual % Low/ Moderate Income Beneficiaries
Van Ham	CFC	$498,505	$2,995	7/1/95	6/30/97	251	282	89.0%			
TDHCA Total Uvalde County		**$498,505**	**$2,995**			**251**	**282**	**89.0%**			

Table B.60
Uvalde County TDHCA Project Activity

	Distribution of Project Activities										
	Water	Sewer	Septic Tank	Planning	Platting	Street Paving	Drainage	Community Center	Housing	Comprehensive Demonstration Project	Other
Number of Colonias Benefiting from Activity	1	1									

Table B.61
Webb County TDHCA Colonia Funding

Colonia Name	Fund Used	Grant Amount	Matching Funds	Start Date	Planned Completion Date	Proposed Low/ Moderate Income Beneficiaries	Proposed Total Beneficiaries	Proposed % Low/ Moderate Income Beneficiaries	Actual Low/ Moderate Income Beneficiaries	Actual Total Benefic-iaries	Actual % Low/ Moderate Income Beneficiaries
Larga Vista	CFP	$35,750	$0	5/1/93	10/31/94	392	439	89.3%	392	439	89.3%
Bruni	CF	$235,420	$97,580	5/1/93	4/30/95	428	537	79.7%	428	537	79.7%
Oilton & Rodriguez Additio	CF	$235,420	$97,580	5/1/93	4/30/95	428	537	79.7%	428	537	79.7%
Larga Vista	CFC	$500,000	$39,000	4/1/94	3/31/96	160	160	100.0%			
Oilton & Rodriguez Additio	CFC	$500,000	$39,000	4/1/94	3/31/96	160	160	100.0%			
Larga Vista	CFD	$1,000,000	$2,118,798	4/1/94	3/31/97	392	439	89.3%			
Oilton & Rodriguez Additio	CFP	$65,000	$0	4/11/94	12/31/95	404	522	77.4%			
Old Milwaukee	CFP	$65,000	$0	4/11/94	12/31/95	404	522	77.4%			
La Presa	CFP	$49,900	$0	5/1/95	4/30/97	161	177	91.0%			
Los Corralitos	CFC	$300,000	$15,000	7/1/95	6/30/97	37	47	78.7%			
TDHCA Total Webb County		**$2,186,070**	**$2,270,378**			**1,974**	**2,321**	**85.0%**	**820**	**976**	**84.0%**

Table B.62
Webb County TDHCA Project Activity

	Distribution of Project Activities										
	Water	**Sewer**	**Septic Tank**	**Planning**	**Platting**	**Street Paving**	**Drainage**	**Community Center**	**Housing**	**Comprehensive Demonstration Project**	**Other**
Number of Colonias Benefiting from Activity	2			2	2	1			2	1	

Table B.63
Willacy County TDHCA Colonia Funding

Colonia Name	Fund Used	Grant Amount	Matching Funds	Start Date	Planned Completion Date	Proposed Low/ Moderate Income Beneficiaries	Proposed Total Beneficiaries	Proposed % Low/ Moderate Income Beneficiaries	Actual Low/ Moderate Income Beneficiaries	Actual Total Benefic-iaries	Actual % Low/ Moderate Income Beneficiaries
La Sara	CF	$93,200	$0	3/4/92	3/3/94	225	225	100.0%	301	301	100.0%
Lyford South	CF	$93,200	$0	3/4/92	3/3/94	225	225	100.0%	301	301	100.0%
Romoville	CF	$93,200	$0	3/4/92	3/3/94	225	225	100.0%	301	301	100.0%
Zapata Ranch	CF	$93,200	$0	3/4/92	3/3/94	225	225	100.0%	301	301	100.0%
Sebastian	CF	$281,976	$0	5/1/93	4/30/95	917	917	100.0%	912	912	100.0%
La Sara	CFC	$361,150	$1,824,704	4/1/94	3/31/96	800	861	92.9%			
Lyford North	CFP	$27,496	$0	6/1/94	5/31/96	233	445	52.4%			
Lyford South	CFP	$27,496	$0	6/1/94	5/31/96	233	445	52.4%			
Lyford North	CFC	$500,000	$0	7/1/95	6/30/97	163	172	94.8%			
TDHCA Total Willacy County		**$1,263,822**	**$1,824,704**			**2,338**	**2,620**	**89.2%**	**1,213**	**1,213**	**100.0%**

Table B.64
Willacy County TDHCA Project Activity

	Distribution of Project Activities										
	Water	Sewer	Septic Tank	Planning	Platting	Street Paving	Drainage	Community Center	Housing	Comprehensive Demonstration Project	Other
Number of Colonias Benefiting from Activity	5	2		2							

Table B.65
Zapata County TDHCA Colonia Funding

Colonia Name	Fund Used	Grant Amount	Matching Funds	Start Date	Planned Completion Date	Proposed Low/ Moderate Income Beneficiaries	Proposed Total Beneficiaries	Proposed % Low/ Moderate Income Beneficiaries	Actual Low/ Moderate Income Beneficiaries	Actual Total Benefic-iaries	Actual % Low/ Moderate Income Beneficiaries
Medina	CFC	$500,000	$50,000	5/1/95	4/30/97	335	335	100.0%			
Medina	CF	$500,000	$25,000	5/1/93	4/30/95	199	214	93.0%	199	214	93.0%
Medina	CF-2	$300,000	$15,000	9/9/92	9/8/94	346	358	96.6%	430	442	97.3%
TDHCA Total Zapata County		**$1,300,000**	**$90,000**			**880**	**907**	**97.0%**	**629**	**656**	**95.9%**

Table B.66
Zapata County TDHCA Project Activity

	Distribution of Project Activities										
	Water	Sewer	Septic Tank	Planning	Platting	Street Paving	Drainage	Community Center	Housing	Comprehensive Demonstration Project	Other
Number of Colonias Benefiting from Activity		1									

Table B.67
Zavala County TDHCA Colonia Funding

Colonia Name	Fund Used	Grant Amount	Matching Funds	Start Date	Planned Completion Date	Proposed Low/ Moderate Income Beneficiaries	Proposed Total Beneficiaries	Proposed % Low/ Moderate Income Beneficiaries	Actual Low/ Moderate Income Beneficiaries	Actual Total Benefic-iaries	Actual % Low/ Moderate Income Beneficiaries
Loma Alta	CF	$500,000	$0	3/4/92	3/3/94	113	113	100.0%	113	113	100.0%
La Pryor	CFP	$46,500	$0	5/1/93	10/31/94	894	1,257	71.1%	894	1,257	71.1%
Chula Vista	CF	$374,921	$20,000	5/1/93	4/30/95	376	376	100.0%	571	571	100.0%
Colonia Del Norte	CFC	$391,230	$20,000	4/1/94	3/31/96	153	156	98.1%	153	156	98.1%
Chula Vista	CFP	$50,500	$0	4/11/94	12/31/95	642	645	99.5%			
Del Norte	CFP	$50,500	$0	4/11/94	12/31/95	642	645	99.5%			
Loma Alta	CFP	$50,500	$0	4/11/94	12/31/95	642	645	99.5%			
Batesville	CFP	$40,250	$0	5/1/95	4/30/97	1,194	1,313	90.9%			
El Cometa	CFC	$450,000	$25,000	5/1/95	4/30/97	482	496	97.2%			
TDHCA Total Zavala County		**$1,853,401**	**$65,000**			**3,854**	**4,356**	**88.5%**	**1,731**	**2,097**	**82.5%**

Table B.68
Zapata County TDHCA Project Activity

	Distribution of Project Activities										
	Water	Sewer	Septic Tank	Planning	Platting	Street Paving	Drainage	Community Center	Housing	Comprehensive Demonstration Project	Other
Number of Colonias Benefiting from Activity	4			5							

Bibliography

Arroyo, Jorge. Unit Chief, Colonias Planning Unit, Local and Regional Assistance Division, Texas Water Development Board, Austin, Texas. Interviewed by Sheila Cavanagh, February 5, 1997.

Bath, C. Richard, Janet M. Transki and Roberto E. Villarreal. "The Politics of Water Allocation in El Paso County Colonias." *Journal of Borderlands Studies*, vol. 9, no. 1 (Spring 1994).

Border Environment Cooperation Commission. *Draft Project List for Public Information*. Ciudad Juarez, Mexico: December 1996.

Cavanagh, Sheila. "Right Purpose, Wrong Tools: NAFTA's Environmental Institutions and Texas Border Infrastructure." Professional Report, Lyndon B. Johnson School of Public Affairs, The University of Texas at Austin, 1996.

Cedillo, Ruth. Director, Community Development, Texas Department of Housing and Community Affairs. Memorandum to Executive Director Larry Paul Manley, Texas Department of Housing and Community Affairs, November 6, 1995.

Holz, Robert K. and Shane Davies. *Third World Texas: Colonias in the Lower Rio Grande Valley*. Working Paper Series, no. 72. Austin, Texas: Lyndon B. Johnson School of Public Affairs, 1989.

International City/County Management Association. *Texas and New Mexico Colonias: Barriers and Incentives for Local Government Involvement*. Washington, D.C., 1995.

Johnson, Curtis. Chief, Facility Needs Section, Planning Division, Texas Water Development Board, Austin, Texas. Interviewed by Gina Briley, February and March 1996.

Killgore, Mark and David Eaton. *NAFTA Handbook for Water Resource Engineers*. Austin: U.S.-Mexican Policy Studies Program, 1995; and New York: American Society of Civil Engineers, 1995.

Lyndon B. Johnson School of Public Affairs. *Colonia Housing and Infrastructure: Current Population and Housing Characteristics, Future Growth, Housing Water and Wastewater Needs*. Policy Research Project Preliminary Report, Austin, Texas, January 1996.

Mendoza, Steve. Engineering Specialist, Economically Distressed Areas Program, Texas Water Development Board, Austin, Texas. Interviewed by Gina Briley, March 1996.

Ramirez, Roy. Economic Development Representative, Economic Development Administration, U.S. Department of Commerce, Austin, Texas. Interviewed by Sheila Cavanagh, March 21, 1996.

Salinas, Exiquio. *The Colonias Factbook: A Survey of Living Conditions in Rural Areas of South and West Texas Border Counties.* Austin, Texas: Texas Department of Human Services, 1988.

Texas Department of Housing and Community Affairs. *State Low-Income Housing Plan and Annual Report.* Austin, Texas, 1995.

Texas Legislature. El Paso Members. *Draft Report from the El Paso Members of the Texas Legislature Regarding H.B. 1001.* August 18, 1995.

Texas Water Development Board. *Colonia Plumbing Loan Program.* Austin, Texas, February 1995. (Pamphlet.)

Texas Water Development Board. *Water for Texas: Water and Wastewater Needs of Colonias in Texas.* Austin, Texas, October 1992.

Texas Water Development Board. *Water and Wastewater Needs of Texas Colonias: 1995 Update.* Austin, Texas, February 1995.

Tovilla, Edgar. Border Environment Cooperation Commission. Facsimile transmission of unpublished materials to Sheila Cavanagh, November 14, 1996.

Tran, Heidi. Engineering Assistant, Economically Distressed Areas Program, Texas Water Development Board, Austin, Texas. Telephone interview by Gina Briley, April 1996.

Valentin, Paco. Rural Economic and Community Development Agency, U.S. Department of Agriculture, Austin, Texas. Interviewed by Sheila Cavanagh, March 21, 1996.

Walthers, Ann. "Texas Colonias: On the Border of Misery." Professional Report, Institute for Latin American Studies, The University of Texas at Austin, 1988.